Negative Quantum Channels

An Introduction to Quantum Maps that are Not Completely Positive

Synthesis Lectures on Quantum Computing

Editor
Marco Lanzagorta, *U.S. Naval Research Labs*
Jeffrey Uhlmann, *University of Missouri-Columbia*

Negative Quantum Channels: An Introduction to Quantum Maps that are Not Completely Positive
James M. McCracken

ISBN: 978-3-031-01389-8   paperback
ISBN: 978-3-031-02517-4   ebook

DOI 10.1007/978-3-031-01734-6

A Publication in the Springer series
SYNTHESIS LECTURES ON QUANTUM COMPUTING

Lecture #7
Series Editors: Marco Lanzagorta, *U.S. Naval Research Labs*
                Jeffrey Uhlmann, *University of Missouri-Columbia*
Series ISSN
Print 1945-9734   Electronic 1945-9726

Negative Quantum Channels

An Introduction to Quantum Maps that are Not Completely Positive

James M. McCracken
George Mason University

SYNTHESIS LECTURES ON QUANTUM COMPUTING #7

&

ABSTRACT

This book is a brief introduction to negative quantum channels, i.e., linear, trace-preserving (and consistent) quantum maps that are not completely positive. The flat and sharp operators are introduced and explained. Complete positivity is presented as a mathematical property, but it is argued that complete positivity is not a physical requirement of all quantum operations. Negativity, a measure of the lack of complete positivity, is proposed as a tool for empirically testing complete positivity assumptions.

KEYWORDS

complete positivity, quantum process tomography, quantum information, quantum channels, open quantum systems

Dedicated to Angel and G for their patience, understanding, and sense of humor—sorry this JINKIES happened.

Contents

Preface

The study of complete positivity in quantum operations has an interesting history. The timeline seems to start in the mid-1950s [93], although it was not until the late 1960s that it started to get significant attention from physicists, primarily through the work of Sudarshan, Kraus,[1] and others. By the early 1970s, the seminal works of Sudarshan [40], Lindblad [58], and Choi [27] helped cement the usefulness of the complete positivity assumption in dealing with open quantum systems.

All assumptions limit the theories that rely on them,[2] and the limiting nature of the complete positivity assumption was recognized very early on. Sudarshan pointed it out in 1978 [39], and Kraus discussed the dangers of applying his assumptions outside of "scattering-type" experiments in one of the sources most often cited in support of the complete positivity assumption [53]. The early 1990s saw an article/response exchange between Pechukas and Alicki that raised awareness of the issue but did little to settle the argument of whether or not the assumption is required [3, 72, 73]. Modern attitudes towards the assumption seem to be quite varied—from statements of the assumption being physically required [4, 14] to the exact opposite [89].

Current attitudes towards this topic are nicely summed up in some of the comments I have received during the drafting of this book. It was pointed out to me that the use of positivity domains as an alternative to the total domain argument for complete positivity (see Section 5.1) might not be "logically and physically more consistent than requesting complete positivity." A different reviewer (who was kind enough to review my work multiple times) took a stronger stance: "Though it is not an ideological discussion about metaphysics, I understand it may appear such and mine a preconceived position. ...the author states that 'the assumption (of complete positivity) is not empirically justified.' As I explained in my previous report, there is plenty of convincing arguments in the literature that, on the contrary, complete positivity is a physically necessary constraint because of the presence of quantum entanglement. It is true that complete positivity imposes constraints, but attributing them to unnecessary mathematical niceties is wrong. ... avoiding these constraints by limiting the action of a preparation map ... is an ad hoc escamotage. Furthermore, the physical evidence of negative maps should be accompanied by the physical evidence of negative probabilities which is exactly what complete positivity avoids." Another reviewer sums things up nicely with "I am sympathetic to the author's goal of convincing others that the CP assumption is not physically appropriate, but ... there is clear-cut evidence supporting that conclusion that is

---

[1]Kraus' most often cited work, [53], is from 1983, but his publication record on this topic stretches back into the 1960s, e.g., [52] and [42].

[2]Assuming frictionless surfaces in a model of a wooden block sliding down a wooden inclined plane can work well until, for example, you need to explain why the bottom of the block is heating up.

missing from the review." The "clear-cut evidence" mentioned by the third reviewer is a reference to the experimental observations of negative channels in [17]. It would seem that the "clear-cut", decisive, evidence provided by the third reviewer is not possible according to the second reviewer. Clearly, complete positivity is still a topic of passionate debate in the field.

My personal attitude towards complete positivity is, hopefully, clear from this manuscript. I do, however, want to stress that this work is meant as an *introduction*. A full book-length treatment of this topic would take several hundred pages and would, given the current rapid pace of the field, probably be out-of-date rather quickly. As such, many relevant topics (and references) are nowhere to be found in the text that follows. These excluded topics are important, and the hope is that this text provides some motivation for diving deeper into the field. For example, there is a connection between a lack of complete positivity and non-Markovianity. This connection is explored in [79] and [64], and it may provide deeper insight into some of the problems presented in this text. The study of non-Markovian quantum stochastic processes is, however, a vast topic on its own, and its intersection with complete positivity cannot be done justice in the space provided here. I encourage the interested reader to explore the references given above. The bibliography at the end of this work is meant to not only support statements made in the text, but also to help guide the interested reader further into the field.

James M. McCracken
June 2014

Acknowledgments

Cesar Rodriguez-Rosario helped me many times as I wrote this book and cannot be thanked enough. I am indebted to him for everything from the content to the form of this work. I would also like to thank James Troupe for his help with the editing, and Marco Lanzagorta for this opportunity.

James M. McCracken
June 2014

CHAPTER 1

Introduction and Definition of Terms

Table 1.1: These are some of the commonly used notations for this work. Most of the non-standard notation is introduced and explained in dedicated sections of this chapter, but this table is a quick reference for the sets and spaces used throughout.

		Meaning
SETS	$\mathbb{C}$	set of all complex numbers
	$\mathbb{R}$	set of all real numbers
	$\mathbb{R}_+$	set of all positive real numbers
	$\mathbb{Z}$	set of all integers
SPACES	$\mathcal{H}^S$	reduced system Hilbert space
	$\mathcal{H}^B$	bath Hilbert space
	$\mathcal{H}^{SB}$	composite system Hilbert space
	$\mathcal{S}(\mathcal{H}^X)$	set of all valid density matrices on the Hilbert space called X
	$\mathcal{B}(\mathcal{H}^X)$	set of all bounded operators on the Hilbert space called X
OPERATIONS	Tr	trace of an operator or matrix
	$\sharp$	assignment of composite system states to reduced system states (see Sec. 1.4.3)
	$\flat$	partial trace over the bath (see Sec. 1.4.2)
	$\mathbf{A}\odot$	"constructor" called A that takes vectors of states to other objects, e.g., matrices (see Sec. 1.5)

1.1 MOTIVATION

A "quantum operation" was originally introduced in 1961 with a discussion of the stochastic dynamics of quantum systems [94]. Quantum operations have become increasingly important as sub-fields of quantum information move towards experimental verification of key concepts and technologies. In quantum information theory, quantum operations are called "quantum channels" and are formally represented by linear, trace-preserving, completely positive maps (commonly abbreviated as "CPTP"). The "completely positive" qualifier in the definition of a quantum channel can be removed. Such channels are called "negative channels" and are the topic of this book. A negative quantum channel (or quantum operation) is mathematically represented by a linear, trace-preserving map with a consistent assignment of reduced system states to composite system states. It will be shown that the complete positivity requirement is not necessary and may, in many cases, not be a valid physical assumption.

We will use the tools of open quantum systems and quantum information. The theory of open quantum systems provides the most general framework for discussions of quantum dynamics, and quantum information theory provides useful concepts that will help make the discussion more lucid. Several concepts from quantum information will be used without modification (such as qubits[1]), but many other concepts will be redefined. All such redefinitions will be discussed as they are presented. The popularity of quantum information plays a major role in our motivation; hence, most of the examples will be taken from sub-fields such as quantum computing and quantum communication.

The study of open quantum systems can be traced back to Von Neumann [100] and originally came about in the study of the thermodynamics of non-equilibrium systems. The field gained popularity in studies of quantum foundations as a way to understand decoherence and related concepts such as the "quantum-classical" divide. Currently, open systems are enjoying yet another resurgence in popularity as the formalism is applied to simple quantum systems used in quantum information theory. The problem of noise in quantum information is considered somewhere between a serious problem and the reason quantum technologies will ultimately fail to be practical (or even experimentally verified in some cases). As such, traditional open system concepts (e.g., spin baths, spin-boson models, Lindblad evolutions, etc.) have become quite important to the study of quantum information.

We will begin with a very brief introduction to open quantum systems from the quantum information frame of reference. The discussion will then turn to complete positivity and its ubiquitousness in the quantum information literature. We will close with discussions of quantum operations that are not completely positive including examples, implications, and utilities.

1.2   OPEN SYSTEMS

An open quantum system is a quantum system which is coupled to another quantum object (usually called the "bath," "environment," or "reservoir"). A simple sketch of such a system is given in Fig. 1.1.

The system of interest (called "system" in the diagram; also-called the "reduced system") is a subsystem of the combined system (called the "composite system"). The composite system is assumed to be closed and follows Hamiltonian dynamics. The interactions between the reduced system and the bath lead to dynamics of the reduced system which include correlations and cou-

---

[1]Qubits are ideal quantum mechanical systems consisting of only two states. The name harkens back to information theory roots, but the concept of a "two level atom" is a well established and very popular theoretical tool in physics.

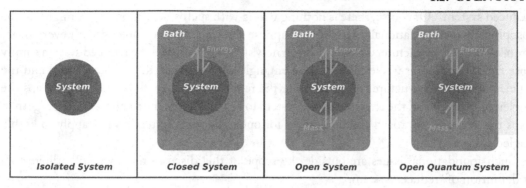

Figure 1.1: An "isolated" system is a system without any kind of bath, a "closed" system has a bath with which it can exchange energy (e.g., heat and work), an "open" system has a bath with which it can exchange both energy and matter, and an "open quantum" system has a bath with which it can exchange energy and matter but with which the boundary is not clearly defined. See the text for further discussion of the system-bath boundary in an open quantum systems. Notice that these definitions apply to non-relativistic systems; see [18] for a discussion of open quantum systems and relativity.

pling between the two subsystems. As a result, the reduced system will, in general, no longer follow unitary Hamiltonian dynamics.

The idea of a quantum system-bath setup is very similar to the traditional ideas of open systems in thermodynamics. The reduced system is free to interact with the environment in any way. Heat, particles, entropy, etc. can all be exchanged across the system-bath interface exactly as in thermodynamics. The difference here is that the system may also become entangled with the bath. This system-bath entanglement will lead to dynamics that are not studied in thermodynamics, and the entanglement with the bath is considered by many to be the primary source of decoherence in experiments [107].

Formal definitions of the reduced system and the bath are disputed. "Open systems" exist in many different fields outside of physics, including computing and social sciences (e.g., [61]), and the general definition is always the same: an "open system" is a system influenced by its environment. In open quantum systems, however, there is a question of system delineation. For example, suppose the reduced system is a single electron and the bath is every other electron in some heterostructure in an experiment. It seems simple enough to call the electron the "reduced system" because it is the focus of the experiment (i.e., it is the "system of interest"). However, this definition introduces the experimenter into the definition of the reduced system. This definition could be rephrased as "the electron is the reduced system because the experimenter wants it to be the reduced system." Of course, there is nothing wrong with such a definition, but it might not be philosophically or foundationally satisfying. Suppose the electron is entangled with every other electron in the heterostructure. Can the electron now still be considered the reduced system simply because the experimenter wishes it? Does the entanglement between the single electron and the sea of electrons in the structure completely blur the reduced system-bath boundary? Perhaps the entanglement requires all the electrons in the system to be considered the reduced system together. Perhaps the standard approach and definitions for open quantum system do not apply to highly entangled systems.

Such foundational issues are outside the scope of this discussion, so we will adhere to a strict definition of "reduced system."

Definition 1.1   The *reduced system* is the part of the experiment directly accessible (e.g., through measurements and preparations) by the experimenter.

See Appendix A.1 for a discussion of this definition. In the example above, the single electron would be called the reduced system if it is the only electron in the system under the control of the experimenter. Naturally, the bath is then defined as the part of the experiment inaccessible to the experimenter. This definition is a straightforward, physical way to divide the open system under investigation. If a qubit is constructed from a double quantum dot system, then the double quantum dot system is the reduced system because the experimenter has a way to control it. The phonon bath that impairs the performance of that qubit is the bath because the experimenter has no way to control it. The experimenter might really want that phonon bath to be part of his reduced system, but he is always restricted by his technological capabilities.[2]

1.3   DENSITY MATRIX

The state of a system will be described by a density matrix rather than a state vector [16, 56, 100]. Suppose the experimenter has some classical uncertainty about the state of a system. Classical uncertainty is the experimenter's ignorance as to the state of the system given several different possible states of the system.[3] For example, he might know the system is in some pure state $|\varphi_i\rangle$ with probability p_i given $\sum_i p_i = 1$. He would then need to take a classical average of quantum expectation values to find the total expectation value for his experiment, i.e.,

$$\langle A \rangle = \sum_i p_i \langle \varphi_i | A | \varphi_i \rangle .$$

---

[2]This restriction is often pointed out by proponents of open systems theories with the platitude "There is no such thing as a closed quantum system."

[3]Classical uncertainty can be contrasted with quantum uncertainty which is a fundamental limit to the precision of simultaneous measurements of conjugate variables of the system (e.g., position and momentum). See [82] for a discussion of the quantum uncertainty principle.

The density matrix is defined as

$$\rho := \sum_i p_i |\varphi_i\rangle\langle\varphi_i| \ .$$

Hence, the expectation value of an operator can be written in terms of the density matrix as

$$\langle A \rangle = \mathrm{Tr}(\rho A) \ .$$

The following properties follow directly from the definition:

$$\mathrm{Tr}(\rho) = 1 \ ,$$

$$\rho^\dagger = \rho \ ,$$

$$\rho \geq 0 \ ,$$

$$\mathrm{Tr}(\rho^2) = 1 \Rightarrow \text{pure state} \ ,$$

where a "pure state" is defined as a state that can be represented by a state vector (i.e., $\rho = |\phi\rangle\langle\phi|$ for some state vector $|\phi\rangle$). The probability of measuring a state ρ in the state $|\phi\rangle$ is the expectation value of the operator $|\phi\rangle\langle\phi|$, i.e.,

$$\begin{aligned} P_\phi &= \mathrm{Tr}(\rho|\phi\rangle\langle\phi|) & (1.1) \\ &= \langle\phi|\rho|\phi\rangle. & (1.2) \end{aligned}$$

Hence, the probability of distinguishing a particular state can be phrased as "measuring the projector" of that particular state.

The positivity requirement for the density matrix

$$\langle\psi|\rho|\psi\rangle \geq 0 \ \forall \ |\psi\rangle \ ,$$

(where $|\psi\rangle$ is some pure state) written as $\rho \geq 0$, insures the eigenvalues of ρ are all positive semi-definite. The unit trace requirement

$$\mathrm{Tr}(\rho) = 1$$

insures all the eigenvalues sum to 1. These two requirements allow the eigenvalues of ρ to be interpreted as formal probabilities, i.e., each eigenvalue is a real number between zero and one, and all the eigenvalues sum to one.

The self-adjoint property for ρ (which is implied by the positivity [31]) is the final lynch pin in the statistical interpretation of the density matrix. All observables in quantum mechanics are represented as linear, self-adjoint operators.

Definition 1.2 A *self-adjoint operator* is an operator that is its own adjoint. The matrix representing a self-adjoint operator is Hermitian [22], i.e., it is equal to its conjugate transpose. We will only be dealing with finite dimensional systems, so self-adjoint operators for us will mean operators with a Hermitian matrix representation.

The state ρ can be shown to be Hermitian, but the self-adjoint requirement extends to all observables. A self-adjoint operator can be decomposed using the spectral theorem into an orthonormal basis, and the operator will have a diagonal matrix representation on this basis (see [68] for a discussion of the spectral decomposition of the density matrix and see [18] for a more complicated discussion of the spectral theorem for operators in the general case). The measurement postulate of quantum mechanics is formulated in terms of projection operators. Projection operators cannot be used to find the off-diagonal matrix elements of the matrix representation of an operator. Notice that if an observable could not be diagonalized in some basis then the operator would always contain some information that was inaccessible to the experimenter by direct observation (i.e., "measurement").

1.4 MATHEMATICAL STRUCTURE OF OPEN SYSTEMS

1.4.1 SYSTEM STATE SPACES

A quantum system in a pure state can be represented by some state vector S in a Hilbert space $\mathcal{H}^S$. A possibly mixed (i.e., not pure) state of the system will be represented by some density operator $\rho^S \in \mathcal{S}(\mathcal{H}^S)$. The set $\mathcal{S}(\mathcal{H}^X)$ is the set of all valid states ρ in the Hilbert space labeled X, and the set $\mathcal{B}(\mathcal{H}^X)$ is the set of all bounded operators in X. This notation will be important in later discussions when it is applied to both the reduced system and the bath Hilbert spaces.

The state of the composite system is defined in a tensor product Hilbert space $\mathcal{H}^S \otimes \mathcal{H}^B$ where $\mathcal{H}^S$ is the Hilbert space of the reduced system and $\mathcal{H}^B$ is the Hilbert space of the bath. The reduced system will have some state

$$\rho^S \in \mathcal{S}(\mathcal{H}^S) \ ,$$

the bath will have some state

$$\rho^B \in \mathcal{S}(\mathcal{H}^B) \ ,$$

and the composite system will have some state

$$\rho^{SB} \in \mathcal{S}(\mathcal{H}^{SB}) \ .$$

The relationship of the composite system to its component states will be important and is referred to as the system-bath correlation. A product state is defined by the following relationship:

$$\rho^{SB} = \rho^S \otimes \rho^B \ ,$$

where $\rho^S \in \mathcal{S}(\mathcal{H}^S)$ and $\rho^B \in \mathcal{S}(\mathcal{H}^B)$. If the state is not a product state, it might still be a separable state, i.e., it might have the form

$$\rho^{SB} = \sum_{ij} p_{ij} \rho_i^S \otimes \rho_j^B \ ,$$

where $\rho_i^S \in \mathcal{S}(\mathcal{H}^S)$, $\rho_j^B \in \mathcal{S}(\mathcal{H}^B)$, $p_{ij} \geq 0 \;\forall i, j$ and $\sum_{ij} p_{ij} = 1$. The set of separable states of the composite system will be denoted $\mathcal{G}^{SB}$. Notice if ρ^{SB} is a product state then $\rho^{SB} \in \mathcal{G}^{SB}$, but the converse is not necessarily true. The composite state is entangled if and only if it is not separable.

1.4.2 PARTIAL TRACE

The relationship between the states of the subsystems and the composite system can be defined by looking at expectation values. The *partial trace* will be defined as the operation that ensures consistent measurement statistics between an observable and its trivial extension to a higher dimensional space.

The trivial extension of an operator is an operator that acts on any extended space as the identity operator. Let $A \in \mathcal{B}(\mathcal{H}^X)$ be an observable with the trivial extension $\tilde{A} = A \otimes I \in \mathcal{B}(\mathcal{H}^{XY}) = \mathcal{B}(\mathcal{H}^X \otimes \mathcal{H}^Y)$ where I is the identity operator. In this example, A acts on space X, so it's trivial extension, $\tilde{A}$, acts on the extended space, Y, as identity.

If $\{|x_i\rangle\}$ is a basis of $\mathcal{H}^X$ and $\{|y_j\rangle\}$ is a basis of $\mathcal{H}^Y$, than $\{|x_i\rangle \otimes |y_j\rangle \equiv |x_i y_j\rangle\}$ is a basis of $\mathcal{H}^{XY}$. A few more definitions are required to completely define this experiment: $\rho^X \in \mathcal{S}(\mathcal{H}^X)$, $\rho^Y \in \mathcal{S}(\mathcal{H}^Y)$, and $\rho^{XY} \in \mathcal{S}(\mathcal{H}^{XY})$ are the states of subsystem X, subsystem Y, and their joint XY system, respectively.

The expectation value of A is given by

$$
\begin{aligned}
\langle A \rangle &= \mathrm{Tr}(\rho^X A) \\
&= \sum_i \langle x_i | \rho^X A | x_i \rangle \quad,
\end{aligned}
$$

and it is assumed that this value should be equal to $\langle \tilde{A} \rangle$. The observable $\tilde{A}$ acts on the joint XY system and is defined as the operation of A on subsystem X and "nothing" on subsystem Y. The joint XY system is defined to have subsystems X and Y, and I is defined as the operator which leaves a system unchanged. Hence, $\langle A \rangle = \langle \tilde{A} \rangle$ is a justified assumption.[4]

The expectation value of $\tilde{A}$ is given by

$$
\begin{aligned}
\langle \tilde{A} \rangle &= \mathrm{Tr}(\rho^{XY} \tilde{A}) \\
&= \sum_{mn} \langle x_m y_n | \rho^{XY} \tilde{A} | x_m y_n \rangle \quad.
\end{aligned}
$$

---

[4]It can be argued that the extension of A to the joint XY system is not $A \otimes I$. It might even be argued that such an extension is not possible [90], i.e., it might be argued that trivial extensions are not valid physical concepts. Such an argument, however, invalidates the definition of the partial trace and has the logical extension of requiring the theorist to take the entire quantum universe into account in the calculation of $\langle A \rangle$.

Inserting the identity operator on the joint XY system, $I^{XY} = \sum_{ij} |x_i y_j\rangle\langle x_i y_j|$, into the above equation yields

$$
\begin{aligned}
\langle \tilde{A} \rangle &= \sum_{mnij} \langle x_m y_n| \rho^{XY} |x_i y_j\rangle\langle x_i y_j| \tilde{A} |x_m y_n\rangle \\
&= \sum_{mnij} \langle x_m y_n| \rho^{XY} |x_i y_j\rangle\langle x_i| A |x_m\rangle \langle y_j|y_n\rangle \\
&= \sum_{mij} \langle x_m y_j| \rho^{XY} |x_i y_j\rangle\langle x_i| A |x_m\rangle \quad .
\end{aligned}
$$

Notice, $\sum_j \langle x_m y_j| \rho^{XY} |x_i y_j\rangle$ are matrix elements of some new state. For convenience, we call this new state $\rho^?$. Notice that $\rho^?$ is defined in terms of ρ^{XY} and is independent of $\mathcal{H}^Y$ because of the sum over the basis $\{y_j\}$. The state $\rho^?$ is literally the result of "tracing out" subsystem Y from the joint system XY, i.e., $\langle x_k| \rho^? |x_l\rangle = \sum_h \langle x_k y_h| \rho^{XY} |x_l y_h\rangle$.

Using this new notation, the expectation value becomes

$$
\begin{aligned}
\langle \tilde{A} \rangle &= \sum_{mi} \langle x_m| \rho^? |x_i\rangle \langle x_i| A |x_m\rangle \\
&= \sum_{m} \langle x_m| \rho^? A |x_m\rangle \\
&= \mathrm{Tr}\left(\rho^? A \right) \quad ,
\end{aligned}
$$

where $\sum_i |x_i\rangle\langle x_i| = I \in \mathcal{B}(\mathcal{H}^X)$ was used.

The original assumption was that

$$
\langle A \rangle = \langle \tilde{A} \rangle \quad ,
$$

which can be rewritten using the above results as

$$
\mathrm{Tr}\left(\rho^X A \right) = \mathrm{Tr}\left(\rho^? A \right) \quad .
$$

From this form of the original assumption, it can be seen that if $\rho^? = \rho^X$, then $\langle A \rangle = \langle \tilde{A} \rangle$.

The matrix elements of the subsystem states (in the $\{|x_i\rangle\}$ basis) are defined in terms of the composite states as

$$
\langle x_k| \rho^X |x_l\rangle = \sum_h \langle x_k y_h| \rho^{XY} |x_l y_h\rangle
$$

and

$$
\langle y_k| \rho^Y |y_l\rangle = \sum_h \langle x_h y_k| \rho^{XY} |x_h y_l\rangle \quad .
$$

The subsystem states are written more succinctly as $\rho^X = \mathrm{Tr}_Y(\rho^{XY})$ and $\rho^Y = \mathrm{Tr}_X(\rho^{XY})$. The operation $\mathrm{Tr}_X(\rho^{XY})$ is called the *partial trace* with respect to X or "tracing out" subsystem X.

This operation is normally introduced in physics texts in the manner it has been presented here [6, 28, 47, 68], but the same partial trace presented above can be defined in a basis-independent fashion [23].

The partial traces preserves the trace of the reduced system, i.e.,

$$\text{Tr}\left(\rho^X\right) = \sum_i \langle x_i | \rho^X | x_i \rangle = \sum_{hi} \langle x_i y_h | \rho^{XY} | x_i y_h \rangle = \text{Tr}\left(\rho^{XY}\right) \ ,$$

where the second to last equality follows from the definition of the matrix elements of subsystem state in terms of the composite state. This result shows that $\text{Tr}\left(\rho^{XY}\right) = 1 \Rightarrow \text{Tr}\left(\rho^X\right) = 1$ if $\rho^X = \text{Tr}_Y\left(\rho^{XY}\right)$. It can also be shown that $\rho^{XY} = \left(\rho^{XY}\right)^\dagger \Rightarrow \rho^X = \left(\rho^X\right)^\dagger$ and $\rho^{XY} \geq 0 \Rightarrow \rho^X \geq 0$ if $\rho^X = \text{Tr}_Y\left(\rho^{XY}\right)$ [71]. Hence, if ρ^{XY} is a valid density operator and ρ^X is defined using the partial trace as $\text{Tr}_Y\left(\rho^{XY}\right)$, then ρ^X will also be a valid density operator.

The partial trace ensures consistency between measurements on the reduced system and the trivial extension of such measurements to the composite system. As such, the introduction of the partial trace leads directly to the definitions of the reduced system and bath states.

Definition 1.3   The state of the reduced system (called the *reduced density matrix* or *reduced state*) is defined as

$$\rho^S := \text{Tr}_B(\rho^{SB}) \ ,$$

and the (much less frequently used) state of the bath is defined as

$$\rho^B := \text{Tr}_S(\rho^{SB}) \ .$$

The partial trace with respect to the bath is a very common operation in the study of open systems, and, for that reason, we will use special notation for tracing out the bath: If $\tau \in \mathcal{S}(\mathcal{H}^{SB})$, then $\tau^\flat \in \mathcal{S}(\mathcal{H}^S)$.

Definition 1.4   The *flat operator* is equivalent to the partial trace with respect to the bath, i.e., $\tau^\flat \equiv \text{Tr}_B(\tau)$.

The reduced state is written down in this notation as

$$\rho^S = (\rho^{SB})^\flat \ .$$

The utility of this notation will become clear in the following sections.

1.4.3   ASSIGNMENTS OF SUBSYSTEM STATES

In general, a complete understanding of the behavior of the reduced system would require knowledge of the composite state. Unfortunately, the reduced system state is defined in terms of the composite state through the use of a non-invertible operation, the partial trace. By definition, the reduced system is the only part of the open system accessible to the experimenter, so how is the experimenter to know ρ^{SB}?

This problem has been addressed by Pechukas and Alicki with the introduction of an assignment map [3, 72, 73]. The assignment map[5] acts as a kind of inverse to the partial trace. For example, define an assignment map A that acts such that if $\tau \in \mathcal{S}(\mathcal{H}^S)$ then

$$A(\tau) \in \mathcal{S}(\mathcal{H}^{SB}) \ .$$

Such a construction leads to some interesting conceptual problems. For example, consider the set of two qubit Bell states

$$|B_0\rangle = \frac{|00\rangle + |11\rangle}{\sqrt{2}} \ ,$$

$$|B_1\rangle = \frac{|00\rangle - |11\rangle}{\sqrt{2}} \ ,$$

$$|B_2\rangle = \frac{|01\rangle + |10\rangle}{\sqrt{2}} \ ,$$

and

$$|B_3\rangle = \frac{|01\rangle - |10\rangle}{\sqrt{2}} \ .$$

One of the two qubits in the above states can be defined as the bath, which leads to

$$(|B_0\rangle\langle B_0|)^{\flat} = (|B_1\rangle\langle B_1|)^{\flat} = (|B_2\rangle\langle B_2|)^{\flat} = (|B_3\rangle\langle B_3|)^{\flat} = \frac{I}{2} \ .$$

Define a reduced state $\tau \in \mathcal{S}(\mathcal{H}^S)$ and composite state $\rho \in \mathcal{S}(\mathcal{H}^{SB})$ such that

$$\rho^{\flat} = \tau \ . \tag{1.3}$$

If $\tau = I/2$, Eq. 1.3 would be satisfied if the composite state ρ was any of the Bell states, and it would not be possible to determine the specific Bell state of ρ with knowledge of τ alone.

The assignment operation could be constructed as

$$A(\tau) = \Gamma \ ,$$

where

$$\Gamma = \{\rho \in \mathcal{S}(\mathcal{H}^{SB}) \mid \rho^{\flat} = \tau\} \ ,$$

---

[5]It should be noted that the term "map" is used throughout the literature, but it is not always clear that the object being discussed is a mathematically well defined map (e.g., see [37] for a basic definition of a map).

i.e., the assignment operator could produce a set of composite states, each of which obeys the expected behavior of Eq. 1.3. Such a construction quickly runs into problems. For example, suppose A is linear and $\tau = \sum_i p_i \tilde{\tau}_i$. The assignment of τ would be

$$A(\tau) = \sum_i p_i A(\tilde{\tau}_i) \; ,$$

but now there is the problem of defining the addition of sets of composite states. Furthermore, the utility of this construction would require some kind of average over a set of composite states.

Discussions of assignment maps in the literature center around discussions of desired mathematical properties. It is not usually stated explicitly, but these discussions also assume the assignment of τ leads to a single state, not a set of states. The other three desired properties of an assignment map defined as $A(\tau) = \rho$ are

- $\rho^\flat = \tau$ (Consistency),

- $\rho \geq 0$ (Positivity), and

- $A(\alpha\tau_1 + \beta\tau_2) = \alpha A(\tau_1) + \beta A(\tau_2)$ (Linearity),

where α and β are complex scalars.[6] These three properties could be defined for an assignment operator that leads to sets of composite states, so it is not always clear in the literature when A is a well-defined map by definition and when it is a general operator with some set of desired properties. We will give assignment operators a very specific definition in hopes of avoiding the confusion that would result otherwise.

Definition 1.5   The *sharp operation* is defined as the operation that assigns a composite state to a reduced system state that has been (or will be) created in the lab (i.e., a reduced system state in the tomography vector of a tomography experiment[7]) in a consistent, linear manner. If $\tau \in \mathcal{S}(\mathcal{H}^S)$ and $\rho \in \mathcal{S}(\mathcal{H}^{SB})$ with $\rho^\flat = \tau$, then

$$\tau^\# = (\rho^\flat)^\# = \rho \; .$$

The sharp operator assigns the reduced system state τ to a single, specific composite state ρ. This operation is consistent by construction, and is defined to be linear:

$$(\alpha\tau_1 + \beta\tau_2)^\# = \alpha\tau_1^\# + \beta\tau_2^\# \; ,$$

where α and β are complex scalars.

---

[6]It will be seen later that tomography requires linearity with respect to complex scalars, not just real scalars.
[7]These ideas will be discussed in Secs. 1.5 and 2.

Notice, however, that the linearity creates problems with the positivity requirement of ρ. Suppose an experimenter creates the following reduced system states in his lab:

$$\begin{aligned}
\tau_0 &= |0\rangle\langle 0| \;, \\
\tau_1 &= |1\rangle\langle 1| \;, \text{ and} \\
\tau_2 &= |+\rangle\langle +| \;,
\end{aligned}$$

where $|+\rangle = 2^{-1/2}(|0\rangle + |1\rangle)$. These three states can be combined to form other reduced system states, such as

$$\tau_3 = \tau_0 + \tau_1 - \tau_2 = |-\rangle\langle -| \;,$$

where $|-\rangle = 2^{-1/2}(|0\rangle - |1\rangle)$. The reduced system state τ_3 is formed as a linear combination of the three states created in the lab, but it is not ever created in the lab itself. It does, however, meet all of the mathematical requirements for a density matrix (see Sec. 1.3); hence, it is a valid reduced system state.

Further suppose that the experimenter knows the behavior of his bath and defines his composite states in the follow way:

$$\begin{aligned}
\tau_0^{\#} &= \tau_0 \otimes \sigma_1\tau_0\sigma_1 = |01\rangle\langle 01| \;, \\
\tau_1^{\#} &= \tau_1 \otimes \sigma_1\tau_1\sigma_1 = |10\rangle\langle 10| \;, \text{ and} \\
\tau_2^{\#} &= \tau_2 \otimes \sigma_1\tau_2\sigma_1 = |++\rangle\langle ++| \;,
\end{aligned}$$

where σ_1 is the standard Pauli operator. For convenience, he might write down his sharp operator as

$$\tau_i^{\#} = \tau_i \otimes \sigma_1\tau_i\sigma_1 \;.$$

The linearity of the sharp operator would then require

$$\tau_3^{\#} = \tau_0^{\#} + \tau_1^{\#} - \tau_2^{\#} = |01\rangle\langle 01| + |10\rangle\langle 10| - |++\rangle\langle ++| \;.$$

Not only does $\tau_3^{\#} \neq \tau_3 \otimes \sigma_1\tau_3\sigma_1$, it is not even a valid composite state (i.e., $\tau_3^{\#}$ is not positive semi-definite). This problem has three possible solutions.

1. The sharp operator is not linear.

2. The sharp operator is not positive.

3. Quantum states cannot be written down as linear combinations of other quantum states.

The last item (#3) is incompatible with quantum mechanics as a linear theory, which we are assuming.[8] Item #3 must, therefore, be discarded. The first solution (#1) is wrong by definition. The sharp operator was defined to be linear in its construction. It, too, must be discarded. The conclusion seems to be that the sharp operator is not a positive operation.

---

[8]For arguments supporting the linearity of quantum mechanics, see [4, 77].

The positivity of reduced system state assignments to composite states is a bit controversial in the literature. Quantum operations are usually defined to be positive, i.e., a quantum operation takes a valid quantum state to a valid quantum state. The positivity of the quantum state is (as discussed previously) an integral part of the statistical interpretation of the density matrix. This point is the end of the story for many authors discussing assignment maps (most famously Alicki [3]): an operation on quantum states that is not positive is not physical.

There is some appeal to this viewpoint. However, it can be shown to be too restrictive. Consider the following theorem due to Pechukas [72].

Theorem 1.6   Pechukas' theorem *states that a linear, consistent, and* positive *sharp operator must take the form*

$$\rho^{\#} = \rho \otimes \beta_0 \ ,$$

where $\rho \in \mathcal{S}(\mathcal{H}^S)$ is any reduced system state and $\beta_0 \in \mathcal{S}(\mathcal{H}^B)$ is a fixed state of the bath.

Pechukas asked the question of what mathematical conditions need to be placed on the assignment map and what those conditions would imply. Pechukas originally only proved his theorem for the qubit case, but general proofs can be found in [48, 77] and [63]. Pechukas' theorem shows that if the sharp operation is required to be linear, positive, and consistent, then it must take a reduced system state to a composite system with a fixed bath.

The positivity of quantum states is required, but notice that the sharp operator is positive on all states that are actually created by the experimentalist in his lab. This point is the key to the construction of the sharp operator. In the example, he prepares three reduced system states. Each of these is represented by a valid reduced system state. He then defines his sharp operator using his knowledge of the bath (perhaps gained from previous experiments or numerical models), but he has only prepared three reduced system states in his lab and has, therefore, only prepared three composite states. His knowledge of the bath is limited to its relationship to the reduced states he actually prepares, and on those states, the sharp operator is positive.

A sharp operator would act as

$$\tau_i^{\#} = \rho_i \ ,$$

where $\tau_i \in \mathcal{L}(\mathcal{H}^S)$ and $\rho_i \in \mathcal{L}(\mathcal{H}^{SB})$ where $\mathcal{L}(\mathcal{H}^X)$ are the set of states in subsystem X prepared in the lab. Notice that preparing reduced system states is equivalent to choosing bath states. Therefore, the experimenter might write down his sharp operator as the consistent and linear operator

$$\tau_i^{\#} = \tau_i \otimes \sigma_1 \tau_i \sigma_1$$

defined (and guaranteed positive) only on $\tau_i \in \mathcal{L}(\mathcal{H}^S)$. In the example above, $\mathcal{L}(\mathcal{H}^S) = \{\tau_0, \tau_1, \tau_2\}$; hence, by assumption, only $\mathcal{L}(\mathcal{H}^{SB}) = \{\tau_0^{\#}, \tau_1^{\#}, \tau_2^{\#}\}$ ever actually exist in the lab. He can define a sharp operation on linear combinations of states in $\mathcal{L}(\mathcal{H}^S)$, but (as shown above) the resulting composite states are not necessarily positive.

Definition 1.7   The *positivity domain* of a quantum operation ε is the set of states $\tau \in \mathcal{S}(\mathcal{H}^S)$ where $\tau = \varepsilon(\rho^S)$ and $\rho^S \in \mathcal{S}(\mathcal{H}^S)$ (we will primarily be concerned with the case where ε is the reduced dynamics, which is defined in the next section).

This definition will be helpful in delineating when the sharp operation yields quantities that do not have clear physical interpretations.

A quantum state represents the experimenter's knowledge of his system. The density matrix representing his quantum state allows him to probabilistically predict the outcomes of measurements he might perform on that state. The bath is defined by its inaccessibility to the experimenter. He does not have full knowledge of the bath; if he did, he could (at least partially) control it through his interactions with the reduced system and the reduced system's interactions with the bath. Hence, the bath is inaccessible to him, in part at least, due to his ignorance of it. It is an interesting philosophical dilemma to demand a valid density matrix representation (which represents the experimenter's knowledge of possible measurement outcomes) of a system defined by the experimenter's ignorance. Why should the composite system by represented by a valid density matrix at all? Should there be a statistical interpretation of the bath, which, by definition, cannot be experimented on directly? These questions point out some of the philosophical difficulties with demanding a positive assignment map.

The sharp operator is a mathematically well-defined tool. At this point, any physical interpretation of the sharp operation is perhaps misguided. A physical interpretation of the sharp operation would be physical justification for something that is essentially an educated guess about the initial state of the bath. The sharp operator is a linear and consistent operation that takes reduced system states prepared in the lab to composite system states. It is a bijection, and it is positive on the reduced system states actually prepared in the lab. It can be extended by linearity to combinations of states, but it is not necessarily positive on these linear combinations. The sharp operation is a tool meant to act as a pseudo-inversion of the non-invertible partial trace. It is nothing more than a tool, but it is crucially important. The relationship of the reduced system and bath states plays a very big role in the reduced system dynamics.

The point of positivity has been belabored here because it will come back to haunt the discussion in later sections. But first, the reduced dynamics will be defined and it will be shown that, for qubit channels, the sharp operation needs to only be defined on four reduced system states.

1.4.4   REDUCED SYSTEM DYNAMICS

The evolution of a pure quantum state $|\Psi\rangle$ is given by Schrödinger's equation:

$$|\dot{\Psi}\rangle = \frac{-i}{\hbar} H |\Psi\rangle \ ,$$

where the "dot" operator is the total time derivative, H is the Hamiltonian that describes the dynamics of the system, and $\hbar = h/2\pi$ where h is Plank's constant. In general, both $|\Psi\rangle$ and H are time dependent. Notice that both $\frac{d}{dt}$ and H are linear operators.

The density operator of a pure state would be written down as $\rho = |\Psi\rangle\langle\Psi|$ and its evolution can be derived as

$$
\begin{aligned}
\dot{\rho} &= \frac{d}{dt}|\Psi\rangle\langle\Psi| \\
&= |\dot{\Psi}\rangle\langle\Psi| + |\Psi\rangle\langle\dot{\Psi}| \\
&= \left(\frac{-i}{\hbar}H|\Psi\rangle\right)\langle\Psi| + |\Psi\rangle\left(\frac{i}{\hbar}\langle\Psi|H^\dagger\right) \\
&= \frac{-i}{\hbar}\left(H|\Psi\rangle\langle\Psi| - |\Psi\rangle\langle\Psi|H\right) \\
&= \frac{-i}{\hbar}[H, \rho] \ ,
\end{aligned}
$$

where $[A, B] = AB - BA$ is the commutator of A and B and $H = H^\dagger$ by definition.[9] This version of Schrödinger's equation is referred to as the Liouville-von Neumann equation.

Notice that if $\rho = \sum_j p_j \tau_j = \sum_j p_j |\psi_j\rangle\langle\psi_j|$, then

$$
\begin{aligned}
\dot{\rho} &= \sum_j p_j \dot{\tau}_j \\
&= \sum_j p_j \frac{-i}{\hbar}\left(H\tau_j - \tau_j H\right) \\
&= \frac{-i}{\hbar}\left(H\left(\sum_j p_j \tau_j\right) - \left(\sum_j p_j \tau_j\right)H\right) \\
&= \frac{-i}{\hbar}\left(H\rho - \rho H\right) \\
&= \frac{-i}{\hbar}[H, \rho] \ .
\end{aligned}
$$

Hence, the von Neumann equation is valid for both pure and mixed states of the system.

The time evolution of the state vector can be thought of in terms of a time-dependent operator U, i.e.,

$$
|\Psi(t)\rangle = U |\Psi(t_0)\rangle \ ,
$$

where $|\Psi(t)\rangle$ is the state vector representing the state of the system at time t and the operator U depends only on times t and t_0.

If the state of the system is represented as a superposition of states at an initial time t_0, then it is expected to still be represented by a superposition at time t. This fundamental concept

---

[9]We will only use standard Hermitian Hamiltonians, but the study of open quantum systems includes many instances of non-Hermitian Hamiltonians.

of quantum mechanics is called the "principle of superposition" of states [56, 66], and is defined as the assumption that if a system is described by a linear combination of states at some initial time, then the time evolution of the system is given by a linear combination of the time evolution of each individual state in that linear combination [56].

For example, if

$$|\Psi(t_0)\rangle = \alpha\,|\psi_1(t_0)\rangle + \beta\,|\psi_2(t_0)\rangle \quad,$$

where α and β are scalars, then it is expected that

$$
\begin{aligned}
|\Psi(t)\rangle &= \alpha\,|\psi_1(t)\rangle + \beta\,|\psi_2(t)\rangle \\
&= U\,|\Psi(t_0)\rangle \\
&= U\left(\alpha\,|\psi_1(t_0)\rangle + \beta\,|\psi_2(t_0)\rangle\right) \quad.
\end{aligned}
$$

This idea leads to the definition of U as a linear operator so that

$$
\begin{aligned}
|\Psi(t)\rangle &= U\left(\alpha\,|\psi_1(t_0)\rangle + \beta\,|\psi_2(t_0)\rangle\right) \\
&= \alpha U\,|\psi_1(t_0)\rangle + \beta U\,|\psi_2(t_0)\rangle \\
&= \alpha\,|\psi_1(t)\rangle + \beta\,|\psi_2(t)\rangle \quad.
\end{aligned}
$$

The linearity of the operator U rests on the assumption that superpositions of states are preserved in time. It is interesting to try to find justification for this assumption within the interpretation of state vectors as representations of the experimenter's knowledge of possible measurement outcomes. Ideas beyond this assumption have been explored in the literature, but such nonlinear theories of quantum mechanics are well beyond the scope of the present work.[10]

The normalization of the state vector to unity is required by the statistical interpretation, which must be valid at all times if it is to be useful. Hence, it is required that

$$
\begin{aligned}
\langle\Psi(t_0)|\Psi(t_0)\rangle &= \langle\Psi(t)|\Psi(t)\rangle \\
&= \langle\Psi(t_0)|\,U^\dagger U\,|\Psi(t_0)\rangle \quad,
\end{aligned}
$$

which is true if $U^\dagger U = I$, i.e., the linear operator U is unitary.

The evolution operator U can be used to rewrite Schrödinger's equation as

$$
\begin{aligned}
\frac{d}{dt}|\Psi(t)\rangle &= \frac{d}{dt}U\,|\Psi(t_0)\rangle \\
&= \frac{-i}{\hbar}HU\,|\Psi(t_0)\rangle \quad,
\end{aligned}
$$

or

$$\frac{d}{dt}U(t,t_0) = \frac{-i}{\hbar}HU(t,t_0) \quad, \tag{1.4}$$

---

[10]It is also interesting to note that some texts (e.g., [84] and [65]) actually claim the reverse of the argument presented here, i.e., superposition states are allowed in quantum theory precisely because Schrödinger evolution is described by a linear differential equation. Empirical evidence agrees very closely with a linear quantum theory, but the reason for that linearity is a philosophical discussion still under debate.

where the dependence of U on the times t and t_0 has been made explicit. Eq. 1.4 can be solved given the initial condition $U(t_0, t_0) = I$ (i.e., the state vector representing the system does not change if time does not change).

A time-independent Hamiltonian makes Eq. 1.4 immediately solvable as

$$U = \exp\left\{\frac{-i(t - t_0)}{\hbar} H\right\} \ , \tag{1.5}$$

and in general [82],

$$U = T \exp\left\{\frac{-i}{\hbar} \int_{t_0}^{t} dt' H(t')\right\} \ ,$$

where the time dependence of H has been made explicit and T is the time-ordering operator, e.g.,

$$T(H(t_1)H(t_2)) = \Theta(t_1 - t_2)H(t_1)H(t_2) + \Theta(t_2 - t_1)H(t_2)H(t_1) \ ,$$

where Θ is the Heaviside operator.

Hence, the operator U can be defined in terms of the Hamiltonian of the system. Most authors like to point out that solutions to Schrödinger's equation (i.e., derivations of U) for time-dependent Hamiltonians are only possible in special, well-studied cases [82]. This work will, for the most part, only be interested in time independent Hamiltonians. As such, Eq. 1.5 will usually be taken as the definition of the system evolution.

Given a density matrix $\rho(t_0)$ at time t_0, the time evolution of the density operator will be shown to be

$$\rho(t) = U\rho(t_0)U^\dagger \ . \tag{1.6}$$

The density operator was originally introduced as a statistical mixture of pure states

$$\rho(t_0) = \sum_i p_i |\phi_i(t_0)\rangle\langle\phi_i(t_0)| \ ,$$

where $p_i \in [0, 1]$, $\sum_i p_i = 1$, and $|\phi_i(t_0)\rangle\langle\phi_i(t_0)|$ is some pure state the system might be in at time t_0. The time evolution of each individual pure state is given above in terms of U, i.e.,

$$|\phi_i(t)\rangle = U |\phi_t(t_0)\rangle \ ,$$

which implies the time evolution of the ith pure state is written down as

$$|\phi_i(t)\rangle\langle\phi_i(t)| = U |\phi_i(t_0)\rangle \left(U |\phi_i(t_0)\rangle\right)^\dagger = U |\phi_i(t_0)\rangle\langle\phi_i(t_0)|U^\dagger \ .$$

The time evolved density operator is written down as

$$\rho(t) = \sum_i p_i |\phi_i(t)\rangle\langle\phi_i(t)| = \sum_i p_i U |\phi_i(t_0)\rangle\langle\phi_i(t_0)|U^\dagger \ ,$$

and the linearity of U immediately implies Eq. 1.6. Also notice that the linearity of U implies that Eq. 1.6 is true for both mixed and pure ρ. In this discussion of the time evolution of ρ, it is implicitly assumed that ρ represents an isolated system.

If a composite system has an initial state ρ^{SB}, then the evolution to the state $\rho^{SB}(t)$ at time t is governed by some composite evolution $U^{SB} \in \mathcal{B}(\mathcal{H}^{SB})$. The initial reduced system state is

$$\rho^S = (\rho^{SB})^\flat \ ,$$

and the reduced system state at time t is

$$\rho^S(t) = \left(U^{SB} \rho^{SB} U^{SB\dagger} \right)^\flat \ .$$

It is assumed that the composite system is isolated and will obey unitary time evolution, but the reduced system dynamics are defined through the flat operator and, in general, will not be unitary. The idea of open system dynamics is to define a more general kind of dynamics for the reduced system by assuming it is part of some larger, isolated, composite system.

Definition 1.8 The *reduced dynamics* (or *reduced system dynamics*) are defined as

$$\rho^S(t) = \left(U^{SB} \left(\rho^S(t_0) \right)^\sharp U^{SB\dagger} \right)^\flat \ , \tag{1.7}$$

where $\rho^S(t) \in \mathcal{S}(\mathcal{H}^S)$ is the reduced system state at time t, $t_0 \leq t$, and $U^{SB} \in \mathcal{B}(\mathcal{H}^{SB})$ is unitary and called the *composite dynamics* (or *composite system dynamics*).

The dynamics of a reduced system initially in the state ρ^S will be described by the map

$$\rho^S(t) = \Phi_t(\rho^S) \equiv \left(U^{SB} (\rho^S)^\sharp U^{SB\dagger} \right)^\flat \ .$$

The map Φ_t is linear, Hermitian, positive (on the positivity domain), and trace preserving. These features are immediately seen from rewriting Φ_t as a composition of three separate maps defined acting on density matrices $\tau \in \mathcal{S}(\mathcal{H}^{SB})$ and $\rho \in \mathcal{S}(\mathcal{H}^S)$ as

$$
\begin{aligned}
\mathsf{U}(\tau) &= U\tau U^\dagger \quad \text{(unitary operator map)} \\
\mathsf{F}(\tau) &= \tau^\flat \quad \text{(flat operator map)} \\
\mathsf{S}(\rho) &= \rho^\sharp \quad \text{(sharp operator map)} \ .
\end{aligned}
$$

The map Φ_t can then be written as

$$\Phi_t(\rho) = (\mathsf{F} \circ \mathsf{U} \circ \mathsf{S})(\rho) = \left(U\rho^\sharp U^\dagger \right)^\flat \ .$$

The maps U, F, and S are all linear, trace preserving, positive (on some domain), and Hermiticity preserving. Hence, Φ_t preserves the statistical interpretation on the density matrix representation of the reduced state in the positivity domain. At any given time, t, to which the reduced system evolves by Φ_t, the reduced system will always be described by a valid density matrix. It is important to emphasize this point because no mention has yet been made of complete positivity (which is another mathematical requirement often imposed on the reduced dynamics described by Φ_t). The "correctness" of the description of the reduced dynamics described by Φ_t is determined by empirical evidence, but Φ_t is mathematically well defined and preserves the statistical interpretation of the reduced system states.

1.5   "VECTOR OF STATES" NOTATION

There is no standard notation in the study of open quantum systems. Some authors prefer index notation while others prefer the more abstract notions of operators on states. Our notation will attempt keep a close connection with the lab through the tomography vectors that might otherwise be lost in the math.

Definition 1.9   A *vector of states* is defined as a row vector of N^2 matrices that form a basis for $N \times N$ matrices.

For example, a vector of states for a qubit channel consists of four states. All the examples in this section will be given for qubit channels for the sake of simplicity.

Definition 1.10   A *tomography vector* is a vector of states where each matrix in the vector is a valid pure state density matrix.

The order of the states in both the vector of states and the tomography vector is important, e.g., the structure of the transformation matrix introduced below will depend on the order of matrices in the vector of states.

If τ_0, τ_1, τ_2, and τ_3 are all valid density matrices and form a basis, then

$$\vec{\tau} = (\tau_0, \tau_1, \tau_2, \tau_3)$$

is a tomography vector. An example of a vector of states that is not a tomography vector is the Pauli vector

$$\vec{\sigma} = (\sigma_0, \sigma_1, \sigma_2, \sigma_3) \ .$$

Both types of vectors will follow all the rules outlined below.

Definition 1.11   A *transformation matrix* is a matrix of complex values that takes one vector of states to another.

For example, the relationship between two vectors of states $\vec{x}$ and $\vec{y}$ might be defined by the transformation matrix $\hat{T}$ as

$$\vec{y} = \vec{x}\hat{T} \ .$$

The vector $\vec{x}$ is a row vector and $\hat{T}$ is a matrix, therefore each element of the new vector $\vec{y}$ is defined as some linear combination of the elements of $\vec{x}$. The transformation matrix $\hat{T}$ always exists because $\vec{y}$ and $\vec{x}$ are bases, i.e., every member of $\vec{x}$ can be written as a linear combination of the members of $\vec{y}$ by definition and vice versa. Similarly, the vector $\vec{x}$ can be written as a linear combination of the members of $\vec{y}$, so there exists some transformation matrix $\hat{T}'$ such that

$$\vec{x} = \vec{y}\hat{T}' \ .$$

This relationship leads to

$$\vec{y} = \vec{x}\hat{T} = \vec{y}\hat{T}'\hat{T} \ ,$$

which implies $\hat{T}'\hat{T} = \hat{T}\hat{T}' = I$ where I is the identity matrix. Therefore, $\hat{T}' = \hat{T}^{-1}$, i.e., the transformation matrix $\hat{T}$ has an inverse.

An operator or map applied to a vector of states acts on each individual element. For example, the trace of $\vec{x}$ would be

$$\mathrm{Tr}(\vec{x}) = (\mathrm{Tr}(x_0), \mathrm{Tr}(x_1), \mathrm{Tr}(x_2), \mathrm{Tr}(x_3)) \ ,$$

and the sharp operator would act as

$$\vec{x}^{\#} = (x_0^{\#}, x_1^{\#}, x_2^{\#}, x_3^{\#}) \ .$$

A matrix O meant to be multiplied by each element of a vector states is acted on each element of the vector individually as

$$O : \vec{x} = (Ox_0, Ox_1, Ox_2, Ox_3)$$

or

$$\vec{x} : O = (x_0 O, x_1 O, x_2 O, x_3 O) \ .$$

Notice that once again, a new vector of states is the result of this operation. For example, conjugation by a unitary matrix U can be performed on each element of the vector as

$$U : \vec{x} : U^{\dagger} = (Ux_0 U^{\dagger}, Ux_1 U^{\dagger}, Ux_2 U^{\dagger}, Ux_3 U^{\dagger}) \ .$$

All of this notation allows the reduced dynamics to be defined on some tomography vector $\{\vec{\tau}\}_i \in \mathcal{S}(\mathcal{H}^S)$, where $\{\vec{\tau}\}_i$ is the ith element of $\vec{\tau}$, as

$$\vec{\tau}(t) = (U : \vec{\tau}^{\#} : U^{\dagger})^{\flat} \ ,$$

with $U \in \mathcal{B}(\mathcal{H}^{SB})$.

The tensor product of two vectors of states yields another vector of states as follows:

$$\begin{aligned} \vec{a} &= \vec{b} \otimes \vec{c} \\ &= (b_1, b_2, b_3, b_4) \otimes (c_1, c_2, c_3, c_4) \\ &= (b_1 \otimes c_1, b_2 \otimes c_2, b_3 \otimes c_3, b_4 \otimes c_4) \\ &= (a_1, a_2, a_3, a_4) \ , \end{aligned}$$

where $a_i = b_i \otimes c_i$. The tensor product of a vector of states with a matrix from the left or right yields another vector of states defined by the tensor product with that matrix on each element of the original vector of states, i.e.,

$$\rho \otimes \vec{\tau} = (\rho \otimes \tau_1, \rho \otimes \tau_2, \rho \otimes \tau_3, \rho \otimes \tau_4)$$

and

$$\vec{\tau} \otimes \rho = (\tau_1 \otimes \rho, \tau_2 \otimes \rho, \tau_3 \otimes \rho, \tau_4 \otimes \rho) \ .$$

Similarly, the addition and subtraction of vectors of states is defined as the addition and subtraction of the individual elements, i.e.,

$$\vec{x} \pm \vec{y} = (x_1 \pm y_1, x_2 \pm y_2, x_3 \pm y_3, x_4 \pm y_4) \ .$$

The dot product of a vector of states with a vector of complex numbers is defined in analogy to the normal dot product between vectors as

$$\vec{c} \cdot \vec{\tau} = \sum_i^4 c_i \tau_i = c_1 \tau_1 + c_2 \tau_2 + c_3 \tau_3 + c_4 \tau_4 \ ,$$

where $\vec{\tau}$ is a qubit vector of states and $\vec{c}$ is a vector of complex numbers.

Definition 1.12   A *superoperator* is a linear operator that takes an operator to another operator. Specifically, we will be using superoperators represented by matrices that take an operator represented by a matrix to another operator represented by a matrix (e.g., to take density operators to other density operators).

It will be shown in the next section that a superoperator representing a quantum channel can be completely characterized in the lab using the states in a tomography vector. Operations acting on a vector of states to yield a new structure (other than another vector of states) are called constructors and will be denoted with boldface and the operator $\odot$.

Definition 1.13   A *constructor* takes a vector of states to something other than a vector of states. Specifically, we will use constructors to take vectors of states to superoperators or specific kinds of matrices.

For example, a superoperator S constructed using the tomography vector $\vec{\tau}$ would be written using some superoperator constructor **S** as

$$S = \mathbf{S} \odot \vec{\tau} \ .$$

The columns of S will be defined by "columnized"[11] versions of linear combinations of the members of $\vec{\tau}$. The superoperator S is a matrix of complex values constructed from rearranged elements

---

[11]The concept of columnizing matrices will be discussed in depth in the next section.

of linear combinations of the members of $\bar{\tau}$. The main idea of a constructor is to "construct" a new structure from a vector of states. Constructors will prove very useful in characterizing channels. The utility of the superoperator constructor $\mathbf{S}$ will be shown in the next section and other constructors will be introduced in the discussion of complete positivity.

CHAPTER 2

Tomography

It is hard to have a conversation of reduced system dynamics (or quantum processes in general) without mentioning tomography. Tomography is a tool for the study of the internal structure of an object. For example, in geophysics, seismic tomography involves the use of data collected from the propagation of two types of waves originating from earthquake epicenters to model the interior of Earth [59], and, in biophysics, electron tomography involves the use of data collected using beams of electrons to model cellular structures [38]. The term tomography in quantum mechanics is used to mean one of several different things, including quantum state tomography, quantum process tomography, and quantum measurement tomography. In each case, the object being modeled is different, but the basic idea is the same.

A quantum state is represented by a density matrix and a quantum measurement is represented as a projector operator. Suppose the basis $\{|\phi_i\rangle\}$ spans some Hilbert space $\mathcal{H}^\phi$. The state $\tau \in \mathcal{S}(\mathcal{H}^\phi)$ can be written as

$$\tau = \sum_{mn} c_{mn} |\phi_m\rangle\langle\phi_n| \ .$$

A set of possible ideal measurements for an experimenter tasked with determining τ is given by $\mathbb{M} = \{|\phi_i\rangle\langle\phi_i|\}$ where

$$\sum_i |\phi_i\rangle\langle\phi_i| = I$$

with I representing the identity operator on $\mathcal{H}^\phi$. Suppose the experimenter decides to perform the measurement $M_{(x)} = |\phi_x\rangle\langle\phi_x|$. The probability of τ being in the state $|\phi_x\rangle$ is given by

$$
\begin{aligned}
P(M_{(x)}|\tau) &= \mathrm{Tr}(M_{(x)}\tau) \\
&= \mathrm{Tr}\left(|\phi_x\rangle\langle\phi_x|\left(\sum_{mn} c_{mn}|\phi_m\rangle\langle\phi_n|\right)\right) \\
&= \mathrm{Tr}\left(\sum_n c_{xn}|\phi_x\rangle\langle\phi_n|\right) \\
&= c_{xx} \ ,
\end{aligned}
\tag{2.1}
$$

as expected by the construction of τ.

The state of the system is no longer described by τ once the measurement is made. The new state of the system is

$$\tau' = \frac{M_{(x)}\tau M_{(x)}}{P(M_{(x)}|\tau)} \ .$$

Notice

$$
\begin{aligned}
P(M_{(x)}|\tau') &= \mathrm{Tr}(M_{(x)}\tau') \\
&= \mathrm{Tr}\left(M_{(x)}\frac{M_{(x)}\tau M_{(x)}}{P(M_{(x)}|\tau)}\right) \\
&= \frac{1}{P(M_{(x)}|\tau)}\,\mathrm{Tr}\left(M_{(x)}\tau\right) \\
&\equiv \frac{P(M_{(x)}|\tau)}{P(M_{(x)}|\tau)} \\
&= 1 \; ,
\end{aligned}
$$

where the cyclic property of the trace was used and $M_{(x)}^2 = M_{(x)}$ was applied twice. The probability of measuring a system in the state defined by the measurement operator after that measurement has already taken place is always one. The measurement process changes the state of the system and makes it impossible to determine the density matrix describing that system completely without repeating the preparation procedure. This feature of quantum measurement requires several copies of τ to perform tomography of τ.

A state is prepared and measured repeatedly, and the ratio of measurement outcomes to the total number of states prepared is the approximated probability of that state being in the pure state represented by the measurement operator. The experimenter can only prepare a finite number of states, so the probability value will always be an approximation of the theoretical value.

Another difficulty comes from the set of possible ideal measurements used by the experimenter. The set $\mathbb{M}$ is a natural choice of measurements for a state in $\mathcal{H}^\phi$, but as shown above, (given many copies of τ) the experimenter can only determine the diagonal terms of τ with measurements from this set; hence, τ cannot be completely defined only in terms of members of $\mathbb{M}$.

Suppose another set of ideal measurements is

$$
\mathbb{M}' = \{|\varphi_j\rangle\langle\varphi_j|\}
$$

with

$$
\sum_j |\varphi_j\rangle\langle\varphi_j| = I \; .
$$

The set $\{|\phi_i\rangle\}$ spans $\mathcal{H}^\phi$, hence each $|\varphi_j\rangle$ can be expanded as

$$
|\varphi_j\rangle = \sum_m a_{jm}\,|\phi_m\rangle \; ,
$$

where a_{jm} are complex coefficients. The members of $\mathbb{M}'$ are superpositions of the basis $\{|\phi_i\rangle\}$ and can be rewritten to yield

$$
\mathbb{M}' = \left\{\left(\sum_m b_{jm}\,|\phi_m\rangle\right)\left(\sum_n b_{jn}^*\,\langle\phi_n|\right)\right\} = \left\{\sum_{mn} a_{jmn}|\phi_m\rangle\langle\phi_n|\right\} \; ,
$$

where $a_{jmn} = b_{jm}b_{jn}^*$. A measurement operator taken from this new set would be

$$M'_{(x)} = |\varphi_x\rangle\langle\varphi_x| = \sum_{mn} a_{xmn}|\phi_m\rangle\langle\phi_n| \; ,$$

and the probability of measurement for $M'_{(x)}$ would be

$$
\begin{aligned}
P(M'_{(x)}|\tau) &= \mathrm{Tr}(M'_{(x)}\tau) \\
&= \mathrm{Tr}\left(\sum_{ij} a_{xij}|\phi_i\rangle\langle\phi_j|\left(\sum_{mn} c_{mn}|\phi_m\rangle\langle\phi_n|\right)\right) \\
&= \mathrm{Tr}\left(\sum_{imn} a_{xim}c_{mn}|\phi_i\rangle\langle\phi_n|\right) \\
&= \sum_{mn} a_{xnm}c_{mn} \; .
\end{aligned}
\tag{2.2}
$$

The probability of measurement P for a given measurement operator M given some state τ can be found experimentally by measuring M repeatedly and using the ratio of successful measurements to total measurement attempts as an approximation of P.[1] The experimenter can then determine τ from his approximation of P. Following the example above, a given diagonal term of τ would be found using Eq. 2.1 as

$$c_{xx} = P(M_{(x)}|\tau) \; ,$$

hence, with $M_{(x)} \equiv |\phi_x\rangle\langle\phi_x|$ and $P(M_{(x)}|\tau) \equiv \mathrm{Tr}(M_{(x)}\tau)$,

$$\tau = \sum_x P(M_{(x)}|\tau)|\phi_x\rangle\langle\phi_x| + \sum_{mn,m\neq n} c_{mn}|\phi_m\rangle\langle\phi_n| \; .$$

A given off-diagonal term of τ can be found using Eq. 2.2.

This algebra is meant to illustrate the basic point of tomography. In general, the experimenter could produce different sets of ideal measurements and repeat the process to determine other off-diagonal terms of τ. This type of iterative process would continue until the correct linear combination of known coefficients (from the measurement operators consisting of states that are superpositions of the original basis states) and approximated probabilities (from measurements of the measurement operators) is found to completely reconstruct τ (within some error tolerance due to the approximated probabilities). This process is finite because τ is finite (or, at least, it is assumed to be in most quantum information experiments). The simplicity of this concept should not mask the experimental difficulty of such tasks. Tomography experiments can be complicated and are difficult to perform for systems of more than a few qubits [46].

---

[1]Remember that the discussion here focuses on ideal quantum measurements which can be regarded a simple yes-no type questions; e.g., Is the state τ given by the measurement operator M, Yes or No?

The present work deals mostly with qubit channels, where an experimenter needs only four measurement operators to completely determine some unknown qubit density matrix. The set of Pauli matrices is a basis for the set of all 2×2 matrices. Define the Pauli vector as

$$\vec{\sigma} = (\sigma_0, \sigma_1, \sigma_2, \sigma_3) = \left(\begin{pmatrix} 1 & 0 \\ 0 & 1 \end{pmatrix}, \begin{pmatrix} 0 & 1 \\ 1 & 0 \end{pmatrix}, \begin{pmatrix} 0 & -i \\ i & 0 \end{pmatrix}, \begin{pmatrix} 1 & 0 \\ 0 & -1 \end{pmatrix} \right) .$$

Any 2×2 matrix A can be written as

$$A = \vec{r} \cdot \vec{\sigma} ,$$

where $\vec{r}$ is a vector of complex numbers. Notice

$$\begin{aligned} \mathrm{Tr}(\sigma_i A) &= r_0 \, \mathrm{Tr}(\sigma_i \sigma_0) + r_1 \, \mathrm{Tr}(\sigma_i \sigma_1) + r_2 \, \mathrm{Tr}(\sigma_i \sigma_2) + r_3 \, \mathrm{Tr}(\sigma_i \sigma_3) \\ &= 2\delta_{i0} r_0 + 2\delta_{i1} r_1 + 2\delta_{i2} r_2 + 2\delta_{i3} r_3 , \end{aligned}$$

where δ_{ij} is the Kronecker delta operator (i.e., $\delta_{ij} = 1 \Leftrightarrow i = j$ and $\delta_{ij} = 0 \Leftrightarrow i \neq j$). It follows that

$$r_i = \frac{\mathrm{Tr}(\sigma_i A)}{2} .$$

The Pauli matrices are not valid density matrices (thus, they are not valid ideal measurement operators), but valid density matrices can be constructed from them. Define the basis $\{|0\rangle, |1\rangle\}$ as the "natural" orthonormal basis for the qubit Hilbert space $\mathcal{H}^Q$. Three separate sets of ideal measurements can be defined as

$$\mathbb{M}_0 \equiv \{\tau_0, \tau_1\} = \{|0\rangle\langle 0|, |1\rangle\langle 1|\} ,$$

$$\mathbb{M}_+ \equiv \{\tau_2, \tau_3\} = \{|+\rangle\langle +|, |-\rangle\langle -|\} ,$$

and

$$\mathbb{M}_i \equiv \{\tau_4, \tau_5\} = \{|+_i\rangle\langle +_i|, |-_i\rangle\langle -_i|\} ,$$

with

$$\begin{aligned} |+\rangle &= \frac{1}{\sqrt{2}} (|0\rangle + |1\rangle) \\ |-\rangle &= \frac{1}{\sqrt{2}} (|0\rangle - |1\rangle) \\ |+_i\rangle &= \frac{1}{\sqrt{2}} (|0\rangle + i\,|1\rangle) \\ |-_i\rangle &= \frac{1}{\sqrt{2}} (|0\rangle - i\,|1\rangle) . \end{aligned}$$

Notice

$$\tau_0 = \frac{\sigma_0 + \sigma_3}{2},$$
$$\tau_1 = \frac{\sigma_0 - \sigma_3}{2},$$
$$\tau_2 = \frac{\sigma_0 + \sigma_1}{2},$$
$$\tau_3 = \frac{\sigma_0 - \sigma_1}{2},$$
$$\tau_4 = \frac{\sigma_0 + \sigma_2}{2}, \text{ and}$$
$$\tau_5 = \frac{\sigma_0 - \sigma_2}{2},$$

are all valid density operators and can be created in the lab. They are all projectors, so they all can be used as measurement operators. Any experimenter can choose one of the sets of measurement operators (i.e., $\mathbb{M}_0$, $\mathbb{M}_+$, or $\mathbb{M}_i$) and obtain σ_0 along with one of the remaining three Pauli operators. The other two Pauli operators can be obtained with a single member from each of the other two sets. For example, suppose the experimenter decides to use the natural set $\mathbb{M}_0$. He can write the Pauli vector in terms of measurement operators as

$$\vec{\sigma} = (\sigma_0, \sigma_1, \sigma_2, \sigma_3) = (\tau_0 + \tau_1, \tau_0 + \tau_1 - 2\tau_3, \tau_0 + \tau_1 - 2\tau_5, \tau_0 - \tau_1) ,$$

which requires only the four operators τ_0, τ_1, τ_3, and τ_5. He could use the same set and write the Pauli vector as

$$\vec{\sigma} = (\tau_0 + \tau_1, 2\tau_2 - \tau_0 - \tau_1, 2\tau_4 - \tau_0 - \tau_1, \tau_0 - \tau_1) ,$$

which requires τ_0, τ_1, τ_2, and τ_4. He might also choose the set $\mathbb{M}_i$ and write it down as

$$\vec{\sigma} = (\tau_4 + \tau_5, 2\tau_2 - \tau_4 - \tau_5, \tau_4 - \tau_5, 2\tau_0 - \tau_4 - \tau_5) ,$$

which requires τ_4, τ_5, τ_2, and τ_0. In this way, the experimenter can express the Pauli vector (i.e., any qubit density operator) with a set of four measurement operators. This set is a tomography vector for a qubit channel, e.g.,

$$\vec{\tau} = (\tau_0, \tau_3, \tau_5, \tau_1) .$$

For reference, every possible tomography vector constructed from members of $\mathbb{M}_0$, $\mathbb{M}_+$, and $\mathbb{M}_i$ has been written in Table 2.1.

A qubit density matrix is a 2×2 matrix, hence, following the discussion above, there is a way to write ρ as

$$\rho = \vec{r} \cdot \vec{\sigma} .$$

The tomography vector $\vec{\tau}$ is also a basis for any qubit density matrix, so ρ can also be written as

$$\rho = \vec{c} \cdot \vec{\tau} ,$$

Table 2.1: These are the possible tomography vectors (up to permutation) for a qubit channel formed using members of the measurement sets described in the text. The vectors are grouped by the set of ideal measurement operators used to form σ_0. See the text for definitions of those sets and the various states.

$\mathbb{M}_0$	$\mathbb{M}_+$	$\mathbb{M}_i$
$(\tau_0, \tau_1, \tau_2, \tau_4)$	$(\tau_2, \tau_3, \tau_4, \tau_0)$	$(\tau_4, \tau_5, \tau_0, \tau_2)$
$(\tau_0, \tau_1, \tau_3, \tau_4)$	$(\tau_2, \tau_3, \tau_5, \tau_0)$	$(\tau_4, \tau_5, \tau_1, \tau_2)$
$(\tau_0, \tau_1, \tau_2, \tau_5)$	$(\tau_2, \tau_3, \tau_4, \tau_1)$	$(\tau_4, \tau_5, \tau_0, \tau_3)$
$(\tau_0, \tau_1, \tau_3, \tau_5)$	$(\tau_2, \tau_3, \tau_5, \tau_1)$	$(\tau_4, \tau_5, \tau_1, \tau_3)$

where $\vec{c}$ is a vector of complex coefficients. These two representations of the state ρ lead to

$$\rho = \vec{r} \cdot \vec{\sigma} = \vec{c} \cdot \vec{\tau} \ .$$

The notation for transforming between $\vec{\sigma}$ and $\vec{\tau}$ would be

$$\vec{\sigma} = \vec{\tau} \hat{R} \ .$$

Example 2.1   Given $\vec{\tau} = (\tau_0, \tau_2, \tau_4, \tau_1)$,

$$\hat{R} = \begin{pmatrix} 1 & -1 & -1 & 1 \\ 0 & 2 & 0 & 0 \\ 0 & 0 & 2 & 0 \\ 1 & -1 & -1 & -1 \end{pmatrix}$$

which leads to

$$\begin{aligned}
\vec{\tau}\hat{R} &= (\tau_0, \tau_2, \tau_4, \tau_1) \begin{pmatrix} 1 & -1 & -1 & 1 \\ 0 & 2 & 0 & 0 \\ 0 & 0 & 2 & 0 \\ 1 & -1 & -1 & -1 \end{pmatrix} \\
&= (\tau_0 + \tau_1, -\tau_0 + 2\tau_2 - \tau_1, -\tau_0 + 2\tau_4 - \tau_1, \tau_0 - \tau_1) \\
&= (\sigma_0, \sigma_1, \sigma_2, \sigma_3) \ .
\end{aligned}$$

In general, the transformation matrix $\hat{R}$ can be found by recognizing that the above vector of states equation is equivalent (by a restacking procedure) to the matrix equation

$$\sigma_M = \tau_M \hat{R}$$

where

$$\sigma_M = (\text{col}(\sigma_0) \ \text{col}(\sigma_1) \ \text{col}(\sigma_2) \ \text{col}(\sigma_3))$$

and

$$\tau_M = \left(\mathrm{col}\left(\{\vec{\tau}\}_0\right) \; \mathrm{col}\left(\{\vec{\tau}\}_1\right) \; \mathrm{col}\left(\{\vec{\tau}\}_2\right) \; \mathrm{col}\left(\{\vec{\tau}\}_3\right) \right)$$

with $\{\vec{\tau}\}_i$ as the ith element of $\vec{\tau}$. The restacking procedure col will be discussed in more detail at the end of this section, but the main idea here is that the vector of state equation for transforming between two vector of states can be written using standard matrices by correctly stacking the elements of each state in the vector of states into a matrix. This new expression implies

$$\hat{R} = \tau_M^{-1}\sigma_M \; .$$

Example 2.2 Following the example above with $\vec{\tau} = (\tau_0, \tau_2, \tau_4, \tau_1)$, this expression yields

$$\hat{R} = \left(\begin{pmatrix} 1 & \frac{1}{2} & \frac{1}{2} & 0 \\ 0 & \frac{1}{2} & \frac{i}{2} & 0 \\ 0 & \frac{1}{2} & -\frac{i}{2} & 0 \\ 0 & \frac{1}{2} & \frac{1}{2} & 1 \end{pmatrix} \right)^{-1} \begin{pmatrix} 1 & 0 & 0 & 1 \\ 0 & 1 & i & 0 \\ 0 & 1 & -i & 0 \\ 1 & 0 & 0 & -1 \end{pmatrix} = \begin{pmatrix} 1 & -1 & -1 & 1 \\ 0 & 2 & 0 & 0 \\ 0 & 0 & 2 & 0 \\ 1 & -1 & -1 & -1 \end{pmatrix} .$$

As explained in Sec. 1.5, $\hat{R}$ has an inverse, therefore

$$\vec{\tau} = \vec{\sigma}\,\hat{R}^{-1} \; .$$

Example 2.3 Following the above example,

$$\begin{aligned}
\vec{\sigma}\,\hat{R}^{-1} &= (\sigma_0, \sigma_1, \sigma_2, \sigma_3) \begin{pmatrix} \frac{1}{2} & \frac{1}{2} & \frac{1}{2} & \frac{1}{2} \\ 0 & \frac{1}{2} & 0 & 0 \\ 0 & 0 & \frac{1}{2} & 0 \\ \frac{1}{2} & 0 & 0 & -\frac{1}{2} \end{pmatrix} \\
&= \left(\frac{\sigma_0 + \sigma_3}{2}, \frac{\sigma_0 + \sigma_1}{2}, \frac{\sigma_0 + \sigma_2}{2}, \frac{\sigma_0 - \sigma_3}{2} \right) \\
&= (\tau_0, \tau_2, \tau_4, \tau_1) \\
&\equiv \vec{\tau} \; .
\end{aligned}$$

This relationship can be useful in moving between the different basis representations of the qubit density matrix, e.g.,

$$\rho = \vec{c} \cdot \vec{\tau} = \vec{r} \cdot \vec{\sigma} = \vec{r} \cdot \left(\vec{\tau}\hat{R} \right) \; .$$

Example 2.4   Again, following the above example, this would lead to

$$
\begin{aligned}
c_0\tau_0 + c_1\tau_2 + c_2\tau_4 + c_3\tau_1 &= r_0\left(\tau_0 + \tau_1\right) + r_1\left(-\tau_0 + 2\tau_2 - \tau_1\right) \\
&\quad + r_2\left(-\tau_0 + 2\tau_4 - \tau_1\right) + r_3\left(\tau_0 - \tau_1\right) \\
&= \left(r_0 - r_1 - r_2 + r_3\right)\tau_0 + 2r_1\tau_2 \\
&\quad + 2r_2\tau_4 + \left(r_0 - r_1 - r_2 - r_3\right)\tau_1 \ .
\end{aligned}
$$

Given two qubit tomography vectors $\vec{x}$ and $\vec{y}$ related by some transformation matrix $\hat{T}$ as

$$
\vec{x} = \vec{y}\hat{T} \ ,
$$

two coefficient vectors $\vec{m}$ and $\vec{n}$ that define the same state ρ as

$$
\rho = \vec{m} \cdot \vec{x} = \vec{n} \cdot \vec{y} \tag{2.3}
$$

can be related by noticing

$$
\begin{aligned}
\vec{m} \cdot \vec{x} &= \vec{x}\vec{m}^T \\
&= \vec{y}\ \hat{T}\vec{m}^T \\
&= \vec{y}\left(\vec{m}\ \hat{T}^T\right)^T \\
&= \vec{y} \cdot \left(\vec{m}\hat{T}^T\right) \\
&= \left(\vec{m}\hat{T}^T\right) \cdot \vec{y} \ ,
\end{aligned}
$$

where $\hat{T}^T$ is the transpose of $\hat{T}$, $(AB)^T = B^T A^T$ by the properties of the transpose, and $\vec{a} \cdot \vec{b} = \vec{a}\vec{b}^T$ for any row vectors $\vec{a}$ and $\vec{b}$. Comparing this result to Eq. 2.3 implies

$$
\vec{n} = \vec{m}\hat{T}^T \ ,
$$

which means the coefficients from above can be related as

$$
\vec{c} = \vec{r}\hat{R}^T \ ,
$$

where $\hat{R}^T$ is the transpose of $\hat{R}$. The coefficients $\vec{r}$ are straightforward to find given ρ as

$$
\vec{r} = \left(\frac{\mathrm{Tr}(\sigma_0\rho)}{2}, \frac{\mathrm{Tr}(\sigma_1\rho)}{2}, \frac{\mathrm{Tr}(\sigma_2\rho)}{2}, \frac{\mathrm{Tr}(\sigma_3\rho)}{2}\right) \ .
$$

Example 2.5 $\vec{c}$ in the example would become

$$
\begin{aligned}
\vec{r}\hat{R}^T &= \left(\frac{\text{Tr}(\sigma_0\rho)}{2}, \frac{\text{Tr}(\sigma_1\rho)}{2}, \frac{\text{Tr}(\sigma_2\rho)}{2}, \frac{\text{Tr}(\sigma_3\rho)}{2}\right)\begin{pmatrix} 1 & 0 & 0 & 1 \\ -1 & 2 & 0 & -1 \\ -1 & 0 & 2 & -1 \\ 1 & 0 & 0 & -1 \end{pmatrix} \\
&= \left(\frac{1}{2}\text{Tr}\left(\sigma_0\rho - \sigma_1\rho - \sigma_2\rho + \sigma_3\rho\right), \frac{2\,\text{Tr}(\sigma_1\rho)}{2},\right. \\
&\qquad \left.\frac{2\,\text{Tr}(\sigma_2\rho)}{2}, \frac{1}{2}\text{Tr}\left(\sigma_0\rho - \sigma_1\rho - \sigma_2\rho - \sigma_3\rho\right)\right) \\
&= \vec{c} \ .
\end{aligned}
$$

These simple exercises of linear algebra will prove to be useful. The relationships between these different representations can be quite useful to the experimentalist. If a state is written as

$$
\rho = \vec{r} \cdot \vec{\sigma} \ ,
$$

then the validity of ρ as a density matrix puts two conditions on the coefficient vector $\vec{r}$. The first coefficient is set by the unit trace condition of ρ as

$$
r_0 = \frac{\text{Tr}(\sigma_0\rho)}{2} = \frac{\text{Tr}(\rho)}{2} = \frac{1}{2} \ ,
$$

and ρ must be positive semi-definite which yields

$$
\frac{1}{2}\left(\text{Tr}(\rho) \pm \sqrt{(\text{Tr}(\rho))^2 - 4|\rho|}\right) = r_0 \pm \sqrt{r_1^2 + r_2^2 + r_3^2} = \frac{1}{2} \pm \sqrt{r_1^2 + r_2^2 + r_3^2} \geq 0 \ ,
$$

where $|\rho|$ is the determinant of ρ. These "valid state conditions" on $\vec{r}$ will impose equivalent conditions on $\vec{c}$ given their relationship through $\hat{R}$, i.e.,

$$
\frac{1}{2}\left(\text{Tr}(\rho) \pm \sqrt{(\text{Tr}(\rho))^2 - 4|\rho|}\right) = \frac{1}{2}\left(1 \pm \sqrt{c_0^2 + c_1^2 + c_2^2 - 2c_0c_3 + c_3^2}\right) \geq 0 \ .
$$

The main idea of tomography is to find a complete description of the qubit channel using only valid states that can be created in the lab. Consider the following example.

Example 2.6 The experimentalist might be looking for the state

$$
\rho = \begin{pmatrix} a & b \\ c & d \end{pmatrix}
$$

that can be written down as

$$\rho = a \begin{pmatrix} 1 & 0 \\ 0 & 0 \end{pmatrix} + b \begin{pmatrix} 0 & 1 \\ 0 & 0 \end{pmatrix} + c \begin{pmatrix} 0 & 0 \\ 1 & 0 \end{pmatrix} + d \begin{pmatrix} 0 & 0 \\ 0 & 1 \end{pmatrix} .$$

The matrices associated with the coefficients a and d are valid density matrices (τ_0 and τ_1, respectively) and can, therefore, be created in the lab. The other two matrices in this basis are clearly not states, but they can be written down in a basis of states by noticing

$$\begin{pmatrix} 0 & 1 \\ 0 & 0 \end{pmatrix} = \frac{\sigma_1 + i\sigma_2}{2} = \vec{r}_b \cdot \vec{\sigma}$$

and

$$\begin{pmatrix} 0 & 0 \\ 1 & 0 \end{pmatrix} = \frac{\sigma_1 - i\sigma_2}{2} = \vec{r}_c \cdot \vec{\sigma} ,$$

with $\vec{r}_b = (0, 1/2, i/2, 0)$ and $\vec{r}_c = (0, 1/2, -i/2, 0)$. The experimenter can find these states in any tomography basis he wishes by simply finding the appropriate transformation matrix $\hat{R}$. Following the above example once more, he would have

$$\begin{aligned}
\vec{c}_b &= \vec{r}_b \hat{R}^T \\
&= (0, 1/2, i/2, 0) \begin{pmatrix} 1 & 0 & 0 & 1 \\ -1 & 2 & 0 & -1 \\ -1 & 0 & 2 & -1 \\ 1 & 0 & 0 & -1 \end{pmatrix} \\
&= \left(\frac{-(1+i)}{2}, 1, i, \frac{-(1+i)}{2} \right)
\end{aligned}$$

and

$$\begin{aligned}
\vec{c}_c &= \vec{r}_c \hat{R}^T \\
&= (0, 1/2, -i/2, 0) \begin{pmatrix} 1 & 0 & 0 & 1 \\ -1 & 2 & 0 & -1 \\ -1 & 0 & 2 & -1 \\ 1 & 0 & 0 & -1 \end{pmatrix} \\
&= \left(\frac{-(1-i)}{2}, 1, -i, \frac{-(1-i)}{2} \right) .
\end{aligned}$$

This linear algebra reveals

$$\begin{pmatrix} 0 & 1 \\ 0 & 0 \end{pmatrix} = \vec{c}_b \cdot \vec{\tau} = \frac{-(1+i)}{2}\tau_0 + \tau_2 + i\tau_4 + \frac{-(1+i)}{2}\tau_1$$

and

$$\begin{pmatrix} 0 & 0 \\ 1 & 0 \end{pmatrix} = \vec{c}_c \cdot \vec{\tau} = \frac{-(1-i)}{2}\tau_0 + \tau_2 - i\tau_4 + \frac{-(1-i)}{2}\tau_1$$

for this example.[2] The experimenter can use this method to find any density operator in terms of whichever $\vec{\tau}$ is easiest for him to implement in his lab. Notice that other transformation matrices can be used to transform between different tomography bases. All these mathematical representations and the transformations between them will prove useful to an experimentalist who is limited in what he can prepare and measure.

If ρ_x is a valid state, then

$$\rho_y = U\rho_x U^\dagger$$

is a valid state given U is some unitary operator. Any state can be expanded in a tomography basis, i.e., $\rho_x = \vec{r} \cdot \vec{x}$ where $\vec{x}$ is some tomography vector and $\vec{r}$ is a vector of complex numbers, hence

$$
\begin{aligned}
\rho_y &= U\left(\vec{r} \cdot \vec{x}\right) U^\dagger \\
&= U\left(r_0 x_0 + r_1 x_1 + r_2 x_2 + r_3 x_3\right) U^\dagger \\
&= r_0 U x_0 U^\dagger + r_1 U x_1 U^\dagger + r_2 U x_2 U^\dagger + r_3 U x_3 U^\dagger \\
&= \vec{r} \cdot \left(U : \vec{x} : U^\dagger\right) \\
&\equiv \vec{r} \cdot \vec{y} \ ,
\end{aligned}
$$

where $\vec{y}$ is some new tomography vector derived from $\vec{x}$. Any unitary conjugation of a tomography vector yields another tomography vector because unitary conjugation of a basis yields another basis.

The concept of a tomography basis can be used in the tomography of a quantum process. The reduced system is in some state at time t given by

$$\rho(t) = \Phi(\rho_0) \ ,$$

where Φ is the map described in Sec. 1.4.4 and $\rho_0 \equiv \rho(0)$ is the initial state of the reduced system. Any density operator can be written in the tomography basis, thus

$$\rho(t) = \Phi(\vec{c} \cdot \vec{\tau}) = \vec{c} \cdot \Phi(\vec{\tau}) \ .$$

Perhaps it is now evident why the linearity of Φ was insisted upon so tenaciously. The linearity of Φ makes the design of tomography experiments straightforward.

Example 2.7   Using the above tomography vector $\vec{\tau}$,

$$\rho(t) = c_0 \Phi(\tau_0) + c_1 \Phi(\tau_2) + c_2 \Phi(\tau_4) + c_3 \Phi(\tau_1) \ .$$

---

[2]The expansion shown in this example is the only one provided in the discussion of quantum process tomography in [68], but the experimenter has many other choices for a tomography basis.

Determining $\Phi(\tau_i)$ is not necessarily simple. It requires an entire state tomography experiment itself, but the entire qubit process can be completely characterized with only four such experiments.

Example 2.8   From the derivation above (following the same example),

$$
\begin{aligned}
\rho(t) &= \Phi(\rho_0) \\
&= \Phi\left(a \begin{pmatrix} 1 & 0 \\ 0 & 0 \end{pmatrix} + b \begin{pmatrix} 0 & 1 \\ 0 & 0 \end{pmatrix} + c \begin{pmatrix} 0 & 0 \\ 1 & 0 \end{pmatrix} + d \begin{pmatrix} 0 & 0 \\ 0 & 1 \end{pmatrix}\right) \\
&= a\Phi(\tau_0) + b\Phi(\vec{c}_b \cdot \vec{\tau}) + c\Phi(\vec{c}_c \cdot \vec{\tau}) + d\Phi(\tau_1) \\
&= a\Phi(\tau_0) + b\vec{c}_b \cdot \Phi(\vec{\tau}) + c\vec{c}_c \cdot \Phi(\vec{\tau}) + d\Phi(\tau_1) \ .
\end{aligned}
$$

This point is an opportune time to discuss some notational issues in the study of open systems. Most of the notation introduced above is not standard. It is, however, quite useful. Typically, the map Φ is represented as a matrix superoperator possessing the same mathematical properties as Φ, i.e.,

$$
\rho(t) = S\rho_0 \ ,
$$

but notice that if the superoperator S in this expression is represented as a matrix, then the density operator ρ would be represented as a column vector. The density operator, however, has already been introduced as a matrix. The stacking operators, formalized by Havel [41], col and mat will be used to avoid unnecessary confusion. In the qubit case, the stacking operations are seen as

$$
\rho_0 = \begin{pmatrix} a & b \\ c & d \end{pmatrix} \Leftrightarrow \mathrm{col}(\rho_0) = \begin{pmatrix} a \\ c \\ b \\ d \end{pmatrix}, \quad \mathrm{mat}(\mathrm{col}(\rho_0)) = \rho_0 \ .
$$

The superoperator notation would be used as

$$
\rho(t) = \Phi(\rho_0) \Rightarrow \mathrm{col}(\rho(t)) = S\,\mathrm{col}(\rho_0) \ .
$$

Example 2.9   If

$$
\rho(t) = \Phi(\rho_0) = U\rho_0 U^\dagger
$$

for some unitary U, then

$$
\mathrm{col}(\rho(t)) = (U^* \otimes U)\,\mathrm{col}(\rho_0) = S\,\mathrm{col}(\rho_0)
$$

with

$$S = U^* \otimes U \ ,$$

where U^* is the complex conjugate of U [41].

The superoperator S is characterized by the process tomography described above.

Example 2.10 Notice the first column of the superoperator can be characterized by its action on the state $\tau_0(t)$ at some initial time $t = 0$, i.e.,

$$
\begin{aligned}
\mathrm{col}(\tau_0(t)) &= S\,\mathrm{col}(\tau_0(0)) \\
&= \begin{pmatrix} s_{00} & s_{01} & s_{02} & s_{03} \\ s_{10} & s_{11} & s_{12} & s_{13} \\ s_{20} & s_{21} & s_{22} & s_{23} \\ s_{30} & s_{31} & s_{32} & s_{33} \end{pmatrix} \begin{pmatrix} 1 \\ 0 \\ 0 \\ 0 \end{pmatrix} \\
&= \begin{pmatrix} s_{00} \\ s_{10} \\ s_{20} \\ s_{30} \end{pmatrix} .
\end{aligned}
$$

In this way, and again using the tomography vector from the example used throughout this section, the columns of S are characterized as

$$\begin{pmatrix} s_{00} \\ s_{10} \\ s_{20} \\ s_{30} \end{pmatrix} = S\,\mathrm{col}(\tau_0(0))$$

$$\begin{pmatrix} s_{01} \\ s_{11} \\ s_{21} \\ s_{31} \end{pmatrix} = S\,\mathrm{col}(\vec{c}_c \cdot \vec{\tau})$$

$$\begin{pmatrix} s_{02} \\ s_{12} \\ s_{22} \\ s_{32} \end{pmatrix} = S\,\mathrm{col}(\vec{c}_b \cdot \vec{\tau})$$

$$\begin{pmatrix} s_{01} \\ s_{11} \\ s_{21} \\ s_{31} \end{pmatrix} = S\,\mathrm{col}(\tau_1(0)) \ ,$$

where S is a superoperator representing some quantum operation implemented in the lab and each of the tomography states in $\vec{\tau}$ represent the state of the reduced system prepared at time

$t = 0$. The experimenter can completely characterize the operation represented by the matrix S by its action on $\vec{\tau}$.

The qubit channel is characterized performing four separate experiments and is represented as a matrix superoperator S or the map Φ.

Process tomography is the method by which the experimenter will probe his channel and is, therefore, a crucial part of quantum information theory. The discussion of complete positivity will hinge on tomography experiments. Complete positivity is a mathematical requirement imposed on Φ (and S) that needs to be verified experimentally through tomography experiments before it can be considered "correct."

Tomography is a mathematical reconstruction of the reduced dynamics and it may be confusing to consider dynamics that lead to non-positive output states. So, before the discussion turns to complete positivity, the requirement of positivity of the reduced dynamics will be discussed.

CHAPTER 3

Non-Positive Reduced Dynamics

Positivity is a crucial part of the density matrix, and it seems troublesome that reduced dynamics (as they have been introduced here) might give rise to non-positive matrices. This concept is confusing enough that a conversation of generalities is best reserved for later. First, this concept of non-positive reduced dynamics will be introduced with a simple example of a one qubit reduced system interacting with a one qubit bath.

3.1 TWO QUBIT SWAP EXAMPLE

Let

$$\vec{\tau} = (|0\rangle\langle 0|, |+\rangle\langle +|, |+_i\rangle\langle +_i|, |1\rangle\langle 1|)$$

and

$$\vec{\tau}^{\#} = (|0+\rangle\langle 0+|, |+1\rangle\langle +1|, |+_i 0\rangle\langle +_i 0|, |1+\rangle\langle 1+|) \ .$$

Now let

$$
\begin{aligned}
\rho &= |-\rangle\langle -| \\
&= |0\rangle\langle 0| + |1\rangle\langle 1| - |+\rangle\langle +| \\
&= \vec{c} \cdot \vec{\tau} \ ,
\end{aligned}
$$

with $\vec{c} = (1, -1, 0, 1)$. Notice ρ is a valid density matrix. The sharp operator is only defined on the tomography vector, so the extension of ρ to the composite system relies on the linearity of the sharp operator as

$$
\begin{aligned}
\rho^{\#} &= \vec{c} \cdot \vec{\tau}^{\#} \\
&= |0+\rangle\langle 0+| + |1+\rangle\langle 1+| - |+1\rangle\langle +1| \\
&= \frac{1}{2}\begin{pmatrix} 1 & 1 & 0 & 0 \\ 1 & 0 & 0 & -1 \\ 0 & 0 & 1 & 1 \\ 0 & -1 & 1 & 0 \end{pmatrix} \ ,
\end{aligned}
$$

which is not a valid density matrix because it is not positive.[1] If the composite evolution is a swap operation, i.e.,

$$U = \begin{pmatrix} 1 & 0 & 0 & 0 \\ 0 & 0 & 1 & 0 \\ 0 & 1 & 0 & 0 \\ 0 & 0 & 0 & 1 \end{pmatrix},$$

then the reduced dynamics yield

$$
\begin{aligned}
\rho(t) &= \vec{c} \cdot \left(U : \vec{\tau}^{\#} : U^{\dagger} \right)^{\flat} \\
&= \vec{c} \cdot \left((|+0\rangle\langle+0|)^{\flat}, (|1+\rangle\langle1+|)^{\flat}, (|0+_i\rangle\langle0+_i|)^{\flat}, (|+1\rangle\langle+1|)^{\flat} \right) \\
&= \vec{c} \cdot (|+\rangle\langle+|, |1\rangle\langle1|, |0\rangle\langle0|, |+\rangle\langle+|) \\
&= |+\rangle\langle+| + |+\rangle\langle+| - |1\rangle\langle1| \\
&= \begin{pmatrix} 1 & 1 \\ 1 & 0 \end{pmatrix},
\end{aligned}
$$

which is not a valid density matrix because, again, it is not positive.[2] This result seems to imply that if an experimenter were to send the state $|-\rangle\langle-|$ through the channel described by U and the sharp operation defined above, then he would get inconsistent measurement statistics. In fact, it is not even clear how the invalid state of $\rho(t)$ should be interpreted.

A superoperator for this channel can be found with the tomography vector $\vec{\tau}$ and the coefficient vectors $\vec{c}_b$ and $\vec{c}_c$ described in the previous section, i.e.,

$$
\begin{aligned}
\left\{ \left(U : \vec{\tau}^{\#} : U^{\dagger} \right)^{\flat} \right\}_0 &= |+\rangle\langle+| = \frac{1}{2}\begin{pmatrix} 1 & 1 \\ 1 & 1 \end{pmatrix}, \\
\vec{c}_c \cdot \left(U : \vec{\tau}^{\#} : U^{\dagger} \right)^{\flat} &= \frac{-(1-i)}{2}|+\rangle\langle+| + |1\rangle\langle1| - i|0\rangle\langle0| + \frac{-(1-i)}{2}|+\rangle\langle+| \\
&= \frac{1}{2}\begin{pmatrix} -(i+1) & i-1 \\ i-1 & i+1 \end{pmatrix}, \\
\vec{c}_b \cdot \left(U : \vec{\tau}^{\#} : U^{\dagger} \right)^{\flat} &= \frac{-(1+i)}{2}|+\rangle\langle+| + |1\rangle\langle1| + i|0\rangle\langle0| + \frac{-(1+i)}{2}|+\rangle\langle+| \\
&= \frac{1}{2}\begin{pmatrix} -(1-i) & -(i+1) \\ -(i+1) & 1-i \end{pmatrix}, \\
\left\{ \left(U : \vec{\tau}^{\#} : U^{\dagger} \right)^{\flat} \right\}_3 &= |+\rangle\langle+| = \frac{1}{2}\begin{pmatrix} 1 & 1 \\ 1 & 1 \end{pmatrix},
\end{aligned}
$$

---

[1]The eigenvalues of $\rho^{\#}$ are $\{1, \frac{-1}{\sqrt{2}}, \frac{1}{\sqrt{2}}, 0\}$.

[2]The eigenvalues of $\rho(t)$ are $\{\frac{1}{2}\left(1 + \sqrt{5}\right), \frac{1}{2}\left(1 - \sqrt{5}\right)\}$.

which leads to a superoperator

$$S = \frac{1}{2} \begin{pmatrix} 1 & -(1-i) & -(i+1) & 1 \\ 1 & -(i+1) & i-1 & 1 \\ 1 & -(i+1) & i-1 & 1 \\ 1 & 1-i & i+1 & 1 \end{pmatrix} .$$

Notice

$$\text{mat}\,(S\,\text{col}(|-\rangle\langle-|)) = \begin{pmatrix} 1 & 1 \\ 1 & 0 \end{pmatrix} ,$$

as expected.

In the Bloch representation [15, 68], the superoperator S will take a general state to

$$\text{mat}\left(S\,\text{col}\left(\frac{\vec{a}\cdot\vec{\sigma}}{2} \right) \right) = \frac{1}{2} \begin{pmatrix} (1-a_1+a_2) & (1-a_1-a_2) \\ (1-a_1-a_2) & (1+a_1-a_2) \end{pmatrix} ,$$

where $\vec{a} = (1, \sin\theta\cos\phi, \sin\theta\sin\phi, \cos\theta)$ and $\vec{\sigma} = (\sigma_0, \sigma_1, \sigma_2, \sigma_3)$ are the standard Pauli operators. The pure state positivity domain can be defined in terms of θ and ϕ as follows:

$$\cos(2\theta)\cos(2\phi) - 2\sin(\theta+\phi) \geq 1 . \tag{3.1}$$

This condition might imply that S is not a complete description of the channel because not all valid pure states are in the positivity domain.

Notice $\rho = |-\rangle\langle-| \Rightarrow \vec{a} = (1, -1, 0, 0)$ leads to Eq. 3.1 yielding $1 \geq \sqrt{5}$, which is false. Hence, S does not yield a valid output state for $|-\rangle\langle-|$. This idea is encapsulated in the idea of positivity domains of the channel. S is not positive everywhere on the reduced system space, but only positive density matrices have a clear physical interpretation in the lab. S is, therefore, said to have a positivity domain, which is defined such that the superoperator S will take every valid initial density matrix in the positivity domain to a valid final density matrix. For this example above, the pure state positivity domain is given by Eq. 3.1.

Suppose the experimenter is determined to define his channel for $|-\rangle\langle-|$. That state is not in the pure state positivity domain defined above, but he can define a different pure state positivity domain using a different tomography vector. He has the option of performing tomography with that state in the tomography vector. For example, he can use the tomography vector

$$\vec{\tau}' = (|-\rangle\langle-|, |+\rangle\langle+|, |+_i\rangle\langle+_i|, |1\rangle\langle1|) ,$$

but he is now forced to extend his knowledge of the bath. Previously, he defined the sharp operation on three of the states in $\vec{\tau}'$, but he now must define the sharp operator on $|-\rangle\langle-|$. Suppose he knows the bath acts the same on orthogonal pairs of states, hence,

$$(\vec{\tau}')^\# = (|-1\rangle\langle-1|, |+1\rangle\langle+1|, |+_i\,0\rangle\langle+_i0|, |1+\rangle\langle1+|) .$$

The composite dynamics are unchanged from above. The superoperator for this tomography experiment, S', is given by the elements of $\vec{s}$ as

$$S' = \frac{1}{2}\begin{pmatrix} -1 & 2i & -2i & 1 \\ -1 & 0 & 0 & 1 \\ -1 & 0 & 0 & 1 \\ 3 & -2i & 2i & 1 \end{pmatrix} .$$

Notice

$$\text{mat}\left(S \, \text{col}(|1\rangle\langle 1|)\right) = \text{mat}\left(S' \, \text{col}(|1\rangle\langle 1|)\right) = |+\rangle\langle +|$$

and

$$\text{mat}\left(S \, \text{col}(|+\rangle\langle +|)\right) = \text{mat}\left(S' \, \text{col}(|+\rangle\langle +|)\right) = |1\rangle\langle 1| \;;$$

So, there is apparently some overlap in the positivity domains of S and S', even though they are not the same superoperator.[3] Notice, however,

$$\text{mat}\left(S' \, \text{col}(|-\rangle\langle -|)\right) = |1\rangle\langle 1| \neq \text{mat}\left(S \, \text{col}(|-\rangle\langle -|)\right) .$$

The state $|-\rangle\langle -|$ is in the positivity domain of S', but not S.

The experimenter can reiterate this process once more. He can choose another tomography vector and define another positivity domain. He knows the states in his tomography vector must be in the positivity domain of his resulting reduced dynamics (by definition of the sharp operator). Suppose he performs tomography on the channel once more using

$$\vec{\tau}'' = (|-_i\rangle\langle -_i|, |+\rangle\langle +|, |+_i\rangle\langle +_i|, |1\rangle\langle 1|) ,$$

and, following his assumption for the bath,

$$(\vec{\tau}'')^{\#} = (|-_i\,0\rangle\langle -_i 0|, |+1\rangle\langle +1|, |+_i\,0\rangle\langle +_i 0|, |1+\rangle\langle 1+|) .$$

The channel acts as

$$\left(U : (\vec{\tau}'')^{\#} : U^{\dagger}\right)^{\flat} = (|0\rangle\langle 0|, |1\rangle\langle 1|, |0\rangle\langle 0|, |+\rangle\langle +|) ,$$

and the superoperator for this tomography experiment S'' is given by the elements of $\vec{s}\,'$ as

$$S'' = \frac{1}{2}\begin{pmatrix} 3 & -2 & -2 & 1 \\ -1 & 0 & 0 & 1 \\ -1 & 0 & 0 & 1 \\ -1 & 2 & 2 & 1 \end{pmatrix} .$$

---

[3]Formally comparing different superoperators is difficult. It is typically accomplished with the diamond norm [2], but the computational complexity of such a calculation is beyond the goal of this discussion. Here, it suffices to point out that S and S' are different by inspection or even $||S - S'||_1 = \text{Tr}(\sqrt{(S - S')(S - S')^{\dagger}}) = 2\sqrt{2} > 0$.

Notice

$$\text{mat}\left(S \,\text{col}(|1\rangle\langle1|)\right) = \text{mat}\left(S' \,\text{col}(|1\rangle\langle1|)\right) = \text{mat}\left(S'' \,\text{col}(|1\rangle\langle1|)\right) = |+\rangle\langle+| \qquad (3.2)$$

and

$$\text{mat}\left(S \,\text{col}(|+\rangle\langle+|)\right) = \text{mat}\left(S' \,\text{col}(|+\rangle\langle+|)\right) = \text{mat}\left(S'' \,\text{col}(|+\rangle\langle+|)\right) = |1\rangle\langle1| \qquad (3.3)$$

as expected because both $|1\rangle\langle1|$ and $|+\rangle\langle+|$ are in all three tomography vectors.

Suppose a given state is in the shared positivity domain, but not in the tomography vectors, of two different superoperators. For example, $\theta = \pi/4$ and $\phi = 0$ describes a state ρ_1, i.e.,

$$\rho_1 = \begin{pmatrix} \frac{1}{2} + \frac{\sqrt{2}}{4} & \frac{\sqrt{2}}{4} \\ \frac{\sqrt{2}}{4} & \frac{1}{2} - \frac{\sqrt{2}}{4} \end{pmatrix},$$

that is in the positivity domain of both S and S'' but not in the tomography vector of either. The output states found using these two superoperators are

$$\text{mat}\left(S \,\text{col}(\rho_1)\right) = \begin{pmatrix} \frac{1}{2} - \frac{\sqrt{2}}{4} & \frac{1}{2} - \frac{\sqrt{2}}{4} \\ \frac{1}{2} - \frac{\sqrt{2}}{4} & \frac{1}{2} + \frac{\sqrt{2}}{4} \end{pmatrix}$$

and

$$\text{mat}\left(S'' \,\text{col}(\rho_1)\right) = \begin{pmatrix} 1 - \frac{\sqrt{2}}{4} & -\frac{\sqrt{2}}{4} \\ -\frac{\sqrt{2}}{4} & \frac{\sqrt{2}}{4} \end{pmatrix}.$$

Notice that

$$\text{mat}\left(S \,\text{col}(\rho_1)\right) = H_d^{-1}\left(\text{mat}\left(S'' \,\text{col}(\rho_1)\right)\right) H_d$$

where

$$H_d = \frac{1}{\sqrt{2}} \begin{pmatrix} 1 & 1 \\ 1 & -1 \end{pmatrix},$$

i.e., the two output states are similar. Consider a second example of $\theta = 3\pi/4$ and $\phi = 0$ which describes a state ρ_2, i.e.,

$$\rho_2 = \begin{pmatrix} \frac{1}{2} - \frac{\sqrt{2}}{4} & \frac{\sqrt{2}}{4} \\ \frac{\sqrt{2}}{4} & \frac{1}{2} + \frac{\sqrt{2}}{4} \end{pmatrix},$$

that is in the positivity domain of both S and S' (not S'') but not in the tomography vector of either. The output states found using these two superoperators are

$$\text{mat}\left(S \,\text{col}(\rho_2)\right) = \text{mat}\left(S \,\text{col}(\rho_1)\right)$$

and

$$\text{mat}\left(S' \,\text{col}(\rho_2)\right) = \begin{pmatrix} \frac{\sqrt{2}}{4} & \frac{\sqrt{2}}{4} \\ \frac{\sqrt{2}}{4} & 1 - \frac{\sqrt{2}}{4} \end{pmatrix}.$$

Again, notice that

$$\text{mat}\left(S\,\text{col}(\rho_2)\right) = H_{dX}^{-1}\left(\text{mat}\left(S'\,\text{col}(\rho_2)\right)\right)H_{dX}$$

where

$$H_{dX} = \sigma_1 H_d \sigma_1 = \frac{1}{\sqrt{2}}\begin{pmatrix} -1 & 1 \\ 1 & 1 \end{pmatrix}\ .$$

So, again, these two output states are similar. It remains an open question as to whether all states in the overlapping positivity domains of different superoperators lead to output states that are similar to each other in the manner seen in the above examples. This point is important to remember when discussing the interpretation of the given representations of negative channels.

3.2 POSITIVITY DOMAINS INTERPRETATION

In the above example, what evidence is there that S, S', and S'' describe the same physical operation? From a theoretical point of view, it might be claimed that the "consistent" construction of the various sharp operations implies identical baths and preparation procedures. A "logical structure" of the bath was provided as follows.

- If the prepared reduced state is in $\{|0\rangle\langle 0|, |1\rangle\langle 1|\}$, then the bath is $|+\rangle\langle +|$.

- If the prepared reduced state is in $\{|+\rangle\langle +|, |-\rangle\langle -|\}$, then the bath is $|1\rangle\langle 1|$.

- If the prepared reduced state is in $\{|+_i\rangle\langle +_i|, |-_i\rangle\langle -_i|\}$, then the bath is $|0\rangle\langle 0|$.

It could be argued that identical bath behavior implies identical channels, i.e., S, S', and S'' are different superoperator representations of the same channel.

The evidence provided from an experimental point of view might be much different. Suppose S, S', and S'' are from three tomography experiments conducted consecutively on some carefully constructed set-up in some carefully controlled environment. It could be argued that the experimentalist has very carefully performed tomography of the same experiment three separate times in three separate ways. "Logic" dictates that he has tomographically characterized the same channel three times. As such, he, like the theorist above, believes S, S', and S'' are different superoperator representations of the same channel.

The interpretation of S, S', and S'' as different superoperator representations of the same channel is difficult to defend. The theorist's argument can be refuted on the grounds of his "logical structure." Ignorance defines the bath. The "theoretical argument" in this example is forced to introduce partial knowledge of the bath, which is suspect.

The "experimental argument" does not fare any better. The three tomography experiments must be made at different times, or must be made on different set-ups. The bath is defined by ignorance, hence claiming identical baths in these three experiments is tantamount to claims of (at least) partial knowledge of the bath. Both the theorist and experimenter must use partial knowledge of the bath to claim S, S', and S'' are different superoperator representations of the same channel.

Partial knowledge of the bath is not necessarily a problem, but it is difficult to justify. The most conservative interpretation is that S, S', and S'' represent different channels. This interpretation implies that every new tomography experiment (leading to a new superoperator) must be considered a new, different channel.

As with everything else in physics, the question of whether or not superoperators with positivity domains are good representations of channels will come from empirical data. In this work, we will assume such superoperators are reasonable and we will show their impact on the complete positivity assumption for quantum operations.

CHAPTER 4

Complete Positivity

This chapter will introduce the concept of complete positivity and the operator representation theorem for completely positive operations. It will be shown that reduced dynamics can always be written as a sum of operators, but only completely positive reduced dynamics can be written as a Kraus operator sum. We will briefly discuss the claim that complete positivity is a physical requirement of nature, and we will point out some problems with this viewpoint. This chapter will conclude with the introduction of the negativity, which will be our primary tool for identifying negative channels.

4.1 COMPLETE POSITIVITY DEFINITION

Mathematically, the requirement of complete positivity can be formulated as follows. Consider an extension of $\mathcal{H}^A$ to the space $\mathcal{H}^A \otimes \mathcal{H}^B$. The map Φ^A is "completely positive" on $\mathcal{H}^A$ if (and only if) $\Phi^A \otimes I_n$ is positive for all such extensions. Following Choi's formulation [26, 27], suppose $\Phi^A(a) \geq 0$ for every $a \geq 0$ and $a \in \mathcal{B}(\mathcal{H}^A)$, then $\Phi^A \otimes I_n$, where I_n is the identity matrix of dimension n, is called "n-positive" if $\Phi^A \otimes I_n \geq 0$. If

$$\Phi^A \otimes I_n \geq 0 \ \forall n \ , \tag{4.1}$$

then Φ^A is called "completely positive." A map that is completely positive is a map that takes a positive operator to another positive operator under any trivial extension. It should be noted that this definition assumes finite spaces, which is made perfectly clear in the math literature (e.g., [27, 93], but can be a little nebulous in the physics literature (e.g., [14]) where discussions of complete positivity with (continuous) infinite baths are often conflated with discussions of Markovity in dynamical semigroup methods. The issues surrounding dynamical semigroup methods will be discussed in Sec. 11.

A fundamental (and very useful) proof due to Choi shows that if $\Phi^A \otimes I_n \geq 0$ when I_n is the same dimension as Φ^A, then Φ^A is completely positive [26, 27]. This result means that complete positivity can be checked just by doubling the dimension of the Hilbert space of the reduced system, which will lead to the construction of a special matrix (called "Choi's matrix") to check for complete positivity.

One of the most important properties of completely positive maps is given by the following theorem.

Theorem 4.1 *The* representation theorem of quantum operations *states that an operation Φ that is:*

- *convex linear, i.e., $\Phi\left(\sum_i p_i \rho_i\right) = \sum_i p_i \Phi_m(\rho_i)$ with $\sum_i p_i = 1$ and $p_i \in [0, 1]$;*

- *completely positive, i.e., $\Phi \otimes I_n \geq 0 \; \forall n$; and*

- *trace preserving, i.e., $\mathrm{Tr}\left(\Phi(\rho)\right) \leq \mathrm{Tr}\left(\rho\right)$*

can be written as

$$\Phi(\rho) = \sum_k E_k \rho E_k^\dagger \tag{4.2}$$

where $\rho \in \mathcal{S}(\mathcal{H}^S)$ and

$$\sum_k E_k^\dagger E_k = I \; .$$

Additionally, any map that can be written in the form of Eq. 4.2 is completely positive, trace-preserving, and convex linear.

Proofs of this theorem can be found in [18, 53], and discussions of this theorem can be found in most standard quantum information textbooks, including [68] and [74]. This operator sum form of a completely positive map (i.e., Eq. 4.2) is referred to as "Choi's form," "Kraus' form" (or, alternatively the "Kraus decomposition" of the operator), or sometimes even "Sudarshan's form."

The Kraus decomposition can be used to describe a completely positive map. However, the practicality of the decomposition is restricted to finite dimensional system-environment Hilbert spaces. For systems with high- (or infinite-) dimensional Hilbert spaces, the Kraus decomposition will lead to an impractically large number of Kraus operators. Examples of infinite dimensional environments occur quite often in the study of thermodynamics and include heat baths and baths of harmonic oscillators (e.g., to model electromagnetic noise in transmission lines). In such cases, completely positive maps are usually dealt with using dynamical semigroup methods. Infinite dimensional baths and semigroup methods will be discussed in Sec. 11.

4.2   REDUCED SYSTEM EVOLUTION AS AN OPERATOR SUM

It is possible to write the reduced dynamics of a system in the Kraus form given some strict assumptions about initial correlations and couplings. The reduced dynamics can be written out explicitly in terms of the spectral forms of the composite state and evolution to show a general form of the reduced dynamics as a sum of operators. Applying specific assumptions to this general form will allow the reduced dynamics to be written in the Kraus form introduced in the previous section.

Suppose the time evolution of the reduced state is due to some interaction with the environment, i.e.,

$$\rho^S(t) = \left(\left(U^{SB}(t)\right) \left(\rho^S\right)^\sharp \left(U^{SB}(t)\right)^\dagger \right)^\flat \; ,$$

where $\left(\rho^S\right)^{\#}$ is the initial state of the composite system and is described (through a spectral decomposition) as

$$\left(\rho^S\right)^{\#} = \sum_i \lambda_i |\Psi_i\rangle\langle\Psi_i| \ ,$$

where each $|\Psi_i\rangle \in \mathcal{H}^{SB}$ can be written in terms of the system and bath basis states, i.e.,

$$|\Psi_i\rangle = \sum_{mn} a_{mn}^{(i)} |s_m b_n\rangle \ .$$

The states $|s_m\rangle \in \mathcal{H}^S$ are an orthonormal basis for the reduced system and $|b_n\rangle \in \mathcal{H}^B$ are an orthonormal basis for the bath. The state of the composite system is then

$$
\begin{aligned}
\left(\rho^S\right)^{\#} &= \sum_i \lambda_i \left(\sum_{mn} a_{mn}^{(i)} |s_m b_n\rangle\right)\left(\sum_{m'n'} a_{m'n'}^{(i)*} \langle s_{m'} b_{n'}|\right) \\
&= \sum_{imnm'n'} \lambda_i a_{mn}^{(i)} a_{m'n'}^{(i)*} |s_m b_n\rangle\langle s_{m'} b_{n'}| \\
&= \sum_{imnm'n'} \lambda_i a_{mn}^{(i)} a_{m'n'}^{(i)*} \left(|s_m\rangle\langle s_{m'}| \otimes |b_n\rangle\langle b_{n'}|\right) \ .
\end{aligned}
$$

The unitary evolution of the composite system can also be written in terms of its spectral decomposition as

$$U^{SB} = \sum_j v_j |\phi_j\rangle\langle\phi_j| \ ,$$

where each combined state of the system and bath can (again) be written in terms of the system and bath basis states, i.e.,

$$|\phi_j\rangle = \sum_{xy} c_{xy}^{(j)} |s_x b_y\rangle \ .$$

The unitary evolution of the composite system[1] can then be written as

$$
\begin{aligned}
U^{SB} &= \sum_{jxyx'y'} v_j c_{xy}^{(j)} c_{x'y'}^{(j)*} |s_x b_y\rangle\langle s_{x'} b_{y'}| \\
&= \sum_{jxyx'y'} v_j c_{xy}^{(j)} c_{x'y'}^{(j)*} \left(|s_x\rangle\langle s_{x'}| \otimes |b_y\rangle\langle b_{y'}|\right) \ .
\end{aligned}
$$

Hence,

$$\left(U^{SB}\right)^{\dagger} = \sum_{okpo'p'} v_k^* c_{op}^{(k)*} c_{o'p'}^{(k)} \left(|s_{o'}\rangle\langle s_o| \otimes |b_{p'}\rangle\langle b_p|\right) \ .$$

[1]Notice that the explicit time dependence of the composite evolution has been dropped from the notation to save space, i.e., $U^{SB}(t) \to U^{SB}$. However, it should be understood that the state of the reduced system at time t will be described by a unitary evolution U^{SB} which, in general, depends on that time t. The complete formal notation of the decomposition of U^{SB} would include explicit time dependence for all of the coefficients, i.e., $v_j \to v_j(t)$ and $c_{xy} \to c_{xy}(t)$.

The reduced system dynamics are then written down as (Appendix A.2)

$$\begin{aligned}
\rho^S(t) &= \left(\left(U^{SB}(t) \right) \left(\rho^S \right)^{\#} \left(U^{SB}(t) \right)^{\dagger} \right)^{\flat} \\
&= \sum_{imnm'n'q} \lambda_i a_{mn}^{(i)} a_{m'n'}^{(i)*} \hat{S}_{qn} |s_m\rangle \langle s_{m'}| \hat{S}_{qn'}^{\dagger} \, ,
\end{aligned}$$

where

$$\begin{aligned}
\hat{S}_{qn} &= \sum_{jxx'} v_j c_{xq}^{(j)} c_{x'n}^{(j)*} |s_x\rangle \langle s_{x'}| && (4.3) \\
&= \left(I \otimes \langle b_q| \right) U^{SB} \left(I \otimes |b_n\rangle \right) && (4.4)
\end{aligned}$$

and

$$\begin{aligned}
\hat{S}_{qn'}^{\dagger} &= \sum_{koo'} v_k^* c_{oq}^{(k)*} c_{o'n'}^{(k)} |s_{o'}\rangle \langle s_o| \\
&= \left(I \otimes \langle b_{n'}| \right) \left(U^{SB} \right)^{\dagger} \left(I \otimes |b_q\rangle \right) \, .
\end{aligned}$$

At this point, the evolution of the reduced system is described as

$$\rho^S(t) = \sum_{imnm'n'q} \lambda_i a_{mn}^{(i)} a_{m'n'}^{(i)*} \hat{S}_{qn} |s_m\rangle \langle s_{m'}| \hat{S}_{qn'}^{\dagger} \, . \qquad (4.5)$$

This sum is not of the form

$$\rho(t) = \sum_q \hat{S}_q \rho \hat{S}_q^{\dagger} \, .$$

Hence, the general evolution of the reduced system is not necessarily completely positive according to the representation theorem of quantum operations. It should be noted that the most general form of reduced dynamics is not only not completely positive but also not linear and not unique [24, 99]. We are, however, assuming linearity throughout for simplicity and to maintain the direct link with standard tomography experiments.

CHAPTER 5

Physical Motivation of Complete Positivity

The argument that all physically reasonable operations are completely positive seems to put strict bounds on the types of operations allowed in nature. As always, rigorous experiments are required to test the complete positivity assumption of quantum operations, but as far as we are aware, no such experiments have yet been conducted.[1] Until such experiments are performed, we are left with a mathematical theory that seems to allow negative channels (as shown above) but argues that such channels are not physically reasonable. There are plenty of examples in theoretical physics of mathematical models being restricted by physically motivated assumptions. So, it is important to understand such arguments for complete positivity.

The current arguments for completely positivity as a physical requirement of all quantum operations can be grouped into two categories which will be called the "total domain argument for complete positivity" and the "product state argument for complete positivity."

5.1 TOTAL DOMAIN ARGUMENT

The thought process behind the total domain argument is as follows: Any known Hilbert space cannot be known to be isolated, i.e., given only $\mathcal{H}^A$, it is not possible to rule out the existence of $\mathcal{H}^{AC} = \mathcal{H}^A \otimes \mathcal{H}^C$ where $\mathcal{H}^C$ is some arbitrary Hilbert space. In fact, it is clear that the system of interest can always be thought of as part of a composite system with a trivial bath. Empirically and mathematically, having no bath is the same has having a trivial bath with respect to the dynamics of the reduced system. In both cases, the dynamics of the reduced system are explained completely in terms of the reduced system.

Suppose ρ is the initial state of a two qubit composite system. Define a state in the composite space as

$$\rho' = I \otimes \varepsilon(\rho) \; . \tag{5.1}$$

It follows that

$$\rho = N \sum_{i,j=0}^{1} |ii\rangle\langle jj| \rightarrow \rho' = N\mathbf{C} \; , \tag{5.2}$$

---

[1]It should be made clear that negative channels have already been observed in the lab [17]. However, without the testing of complete positivity assumptions being the focus point of the experiment, such observations are not considered "rigorous" for our purposes; i.e., they do not provide rigorous validation of a given theory of negative channels.

where N is the appropriate normalization factor. If ρ' must be a valid quantum state for any ρ, then $\rho' \geq 0$ which implies $\mathbf{C} \geq 0$. It will be shown later (see Sec. 6.5) that $\mathbf{C} \geq 0$ implies ε must be completely positive. This is the total domain argument for complete positivity. It can be seen from Eq. 5.2 why this argument has led to the interpretation of complete positivity as a requirement due to possible entanglement between bipartite subsystems.

The total domain argument for complete positivity seems reasonable at first glance. If a map is a valid quantum map, then it must take valid density matrices to valid density matrices. A trivial extension of the map is physically reasonable and must result in a valid quantum map, i.e., the trivial extension of the quantum map must also take valid density matrices to valid density matrices. Therefore, the quantum map must be completely positive. But, notice that the positivity domain is the domain of states in which a map Γ will be positive. On the positivity domain, Γ will take valid initial states to valid final states. Such a requirement is identical in spirit to the requirement of complete positivity except that it is not extended to states which are not actually created in the lab.[2]

5.2 PRODUCT STATE ARGUMENT

Complete positivity can be imposed on physical grounds without using the total domain argument. Some authors instead choose to argue that the reduced system and bath must initially be uncorrelated. Such a situation will always lead to completely positive reduced dynamics.

Initial composite states that have no correlation between the reduced system and the bath can be written as a product state, i.e.,

$$(\rho^S)^\# = \rho^S \otimes \rho^B \ ,$$

where $\rho^S \in \mathcal{S}(\mathcal{H}^S)$ is the initial state of the reduced system and $\rho^B \in \mathcal{S}(\mathcal{H}^B)$ is the initial state of the bath. It will be shown in Sec. 6.2 that the reduced dynamics are completely positive if the initial composite state is a product state.

Originally, Kraus used a product state argument defense for the complete positivity assumption in his work, but it should be noted that he was careful to point out that such assumptions may not hold outside of the scattering-type experiments he was considering [53].

In the quantum information literature, the product state argument is often stated as the necessity of preparation in quantum experiments. It is true that the reduced system must be prepared and ideal preparations destroy all correlations between the reduced system and the bath. However, the reduced dynamics will only be completely positive if the preparation procedure leaves the bath in the same state after every preparation of the reduced system over the course of the entire experiment.

It is not reasonable to assume the preparation procedure has such a feature without evidence. Recovering the complete positivity requirement with an argument of there being some

---

[2]This idea is why the tomography vector is used to define the sharp operation introduced in Sec. 1.4.3.

past composite product state destroys the philosophical advantages of complete positivity. It is equivalent to the remark "If the dynamics are not completely positive, then you've picked the wrong initial state. Your composite system had an earlier composite state that was uncorrelated." How is the experimenter to know when such a time existed for the system? What evidence does he have that such a time ever existed? The initial state of the composite system is the state of the composite system when the experiment begins; any other definition becomes too difficult to justify.

Methods for dealing with negative (i.e., not completely positive) dynamics in the lab will be discussed in later sections, but first, there will be some discussion of how to determine complete positivity.

CHAPTER 6

Measures of Complete Positivity

The operator sum representation theorem is the key to the mathematical convenience of completely positive dynamics, and it can be used to prove a few simple conditions. If the composite dynamics are described by "local unitaries" or if the initial composite state has "zero discord," then completely positive dynamics are guaranteed. These proofs are shown below. Other cases require directly testing the system for complete positivity.

6.1   LOCAL UNITARY COMPOSITE EVOLUTION

The term "locally unitary" composite evolution refers to a composite unitary evolution that is of the form

$$U^{SB} = U^S \otimes U^B \ ,$$

where $U^S \in \mathcal{B}(\mathcal{H}^S)$ and $U^B \in \mathcal{B}(\mathcal{H}^B)$. The reduced dynamics become

$$
\begin{aligned}
\rho^S(t) &= \left(U^{SB} \left(\rho^S \right)^\# \left(U^{SB} \right)^\dagger \right)^\flat \\
&= \left((U^S \otimes U^B) \left(\sum_{imnlk} \lambda_i a_{mn}^{(i)} a_{lk}^{(i)*} |s_m\rangle\langle s_l| \otimes |b_n\rangle\langle b_k| \right) \left(U^{S\dagger} \otimes U^{B\dagger} \right) \right)^\flat \\
&= \sum_{imnlk} \lambda_i a_{mn}^{(i)} a_{lk}^{(i)*} \operatorname{Tr} \left(U^B |b_n\rangle\langle b_k| U^{B\dagger} \right) U^S |s_m\rangle\langle s_l| U^{S\dagger} \\
&= \sum_{imlk} \lambda_i a_{mk}^{(i)} a_{lk}^{(i)*} U^S |s_m\rangle\langle s_l| U^{S\dagger} \\
&= U^S \rho^S U^{S\dagger} \ ,
\end{aligned}
$$

where $\{|s_i\rangle\}$ is an orthonormal basis that spans $\mathcal{H}^S$, $\{|b_i\rangle\}$ is an orthonormal basis that spans $\mathcal{H}^B$,

$$\operatorname{Tr} \left(U^B |b_n\rangle\langle b_k| U^{B\dagger} \right) = \operatorname{Tr} \left(\langle b_n| U^{B\dagger} U^B |b_k\rangle \right) = \operatorname{Tr} \left(\langle b_n|b_k\rangle \right) = \delta_{nk}$$

by the cyclic property of the trace and the unitarity of U^B, $\lambda_i \in \mathbb{R} \ \forall i$, $\{a_{mn}^{(i)}, a_{lk}^{(i)}\} \in \mathbb{C} \ \forall i, m, n, l, k$, and

$$\rho^S = \left(\left(\rho^S \right)^\# \right)^\flat = \left(\sum_{imnlk} \lambda_i a_{mn}^{(i)} a_{lk}^{(i)*} |s_m\rangle\langle s_l| \otimes |b_n\rangle\langle b_k| \right)^\flat = \sum_{iml} \lambda'_{iml} |s_m\rangle\langle s_l| \ ,$$

with

$$\lambda'_{iml} = \lambda_i \sum_k a^{(i)}_{mk} a^{(i)*}_{lk} \ .$$

Any local unitary composite evolution will be completely positive. Notice that this condition applies to the trivial cases of $U^{SB} = U^S \otimes I$, $U^{SB} = I \otimes U^B$, and $U^{SB} = I \otimes I$ where I is the identity operator.

The key idea is that local unitary composite evolution will lead to completely positive reduced dynamics independent of the form of the initial composite state.

6.2 PRODUCT COMPOSITE STATES

Initial composite states that have no correlation between the reduced system and the bath can be written as a product state, i.e.,

$$(\rho^S)^\sharp = \rho^S \otimes \rho^B \ ,$$

where $\rho^S \in \mathcal{S}(\mathcal{H}^S)$ is the initial state of the reduced system and $\rho^B \in \mathcal{S}(\mathcal{H}^B)$ is the initial state of the bath. In this situation, the reduced dynamics become

$$
\begin{aligned}
\rho^S(t) &= \left(U^{SB} (\rho^S)^\sharp \left(U^{SB} \right)^\dagger \right)^\flat \\
&= \left(U^{SB} \left(\rho^S \otimes \rho^B \right) \left(U^{SB} \right)^\dagger \right)^\flat \\
&= \left(U^{SB} \left(\rho^S \otimes \sum_i \lambda_i |\psi_i\rangle\langle\psi_i| \right) \left(U^{SB} \right)^\dagger \right)^\flat \\
&= \sum_{qi} \lambda_i \hat{S}_{qi} \rho^S \hat{S}^\dagger_{qi} \\
&= \sum_j E_j \rho^S E^\dagger_j \ ,
\end{aligned}
$$

where $\rho^B = \sum_i \lambda_i |\psi_i\rangle\langle\psi_i|$ is the spectral decomposition of the (fixed) state of the bath, the operators $\hat{S}_{qi}$ are defined in Eq. 4.3, and the operators $E_j = \sqrt{\lambda_i} \hat{S}_{qi}$ with the index j simply relabeling the double index qi. The reduced system and bath are both classically and quantumly uncorrelated in the above equation, but it is possible to have some correlation in the initial composite state that will still consistently lead to completely positive reduced dynamics (see the next subsection).

One of the simplest ways to force complete positivity in Eq. 4.5 is to make the initial state of the composite system a product state with a fixed bath. To that end, assume the system and the bath are initially uncorrelated, and the bath is in an initial state of $|b_\phi\rangle$. In the derivation of Eq. 4.5, this assumption requires $n = n' = \phi$. The initial state of the reduced system can be found using the decomposition of the composite system as (Appendix A.3)

$$\rho^S = ((\rho^S)^\sharp)^\flat = \sum_{imm'} \lambda_i a^{(i)}_{m\phi} a^{(i)*}_{m'\phi} |s_m\rangle\langle s_{m'}| \ ,$$

and the evolution of the reduced system can now be written as

$$\rho^S(t) = \sum_q \hat{S}_{q\phi} \rho^S \hat{S}_{q\phi}^\dagger \; .$$

The familiar form of the operator sum representation can be recovered with the simple substitution of $\hat{S}_{q\phi} = E_w$, which leads to

$$\rho^S(t) = \sum_w E_w \rho^S E_w^\dagger \; .$$

Notice that it can be shown explicitly from the definition of $\hat{S}_{q\phi}$ (Appendix A.4) that

$$\sum_q \hat{S}_{q\phi}^\dagger \hat{S}_{q\phi} = I \; ,$$

as required by the unit trace requirement on the density matrix, i.e.,

$$\mathrm{Tr}(\rho^S(t)) \;=\; 1 \tag{6.1}$$

$$=\; \mathrm{Tr}\left(\sum_q \hat{S}_{q\phi} \rho^S \hat{S}_{q\phi}^\dagger \right) \tag{6.2}$$

$$=\; \mathrm{Tr}\left(\sum_q \hat{S}_{q\phi}^\dagger \hat{S}_{q\phi} \rho^S \right) \; , \tag{6.3}$$

where, by definition, $\mathrm{Tr}(\rho^S) = 1$.

6.3 ZERO DISCORD INITIAL COMPOSITE STATES

The initial correlations between the reduced system and the bath can be formulated in a information-theoretical sense using concepts such as the classical and quantum mutual information.[1]

Definition 6.1 A *zero discord* state is a state with only classical correlations. It is enough to note here that a state with only classical correlations between the subsystems of $\mathcal{H}^X$ and $\mathcal{H}^Y$ can be written in the form [76]

$$\rho^{XY} = \sum_i \left(\Pi_i^X \otimes I \right) \rho^{XY} \left(\Pi_i^X \otimes I \right) \; ,$$

where $\Pi^X \in \mathcal{B}(\mathcal{H}^X)$ is a projector (i.e., $\left(\Pi^X\right)^2 = \Pi^X$, $\mathrm{Tr}(\Pi^X) = 1$, and $\mathrm{Tr}(\Pi^X A) = \langle\phi| A |\phi\rangle$ with $\Pi^X = |\phi\rangle\langle\phi|$).

---

[1] A discussion of these concepts would be a little off-topic here, but a good introduction can be found in [69, 77, 106, 107] and [78].

The state ρ^{XY} will contain no entanglement between the subsystems $\mathcal{H}^X$ and $\mathcal{H}^Y$, only classical correlations.[2] This form of the initial composite state would be written as [76]

$$(\rho^S)^{\sharp} = \sum_i \left(\Pi_i^S \otimes I\right) \rho^{SB} \left(\Pi_i^S \otimes I\right) = \sum_i \lambda_i \Pi_i^S \otimes \rho_i^B \ ,$$

with $\lambda_i \geq 0 \ \forall i$, $\sum_i \lambda_i = 1$, $\Pi_i^S = |s_i\rangle\langle s_i|$ where $\{|s_i\rangle\}$ is an orthonormal basis of $\mathcal{H}^S$, and ρ_i^B is a valid density operator in $\mathcal{S}(\mathcal{H}^B)$.

This initial composite state will lead to completely positive reduced dynamics. Consider

$$
\begin{aligned}
\rho^S(t) &= \left(U^{SB}(\rho^S)^{\sharp}(U^{SB})^{\dagger}\right)^{\flat} \\
&= \left(U^{SB}\left(\sum_i \lambda_i \Pi_i^S \otimes \rho_i^B\right)(U^{SB})^{\dagger}\right)^{\flat} \\
&= \sum_{qi} \lambda_i \hat{S}_{qi} \Pi_i^S \hat{S}_{qi}^{\dagger} \ ,
\end{aligned}
$$

where, again, the operators $\hat{S}_{qi}$ are defined in Eq. 4.3. Notice that these operators can be expanded as

$$\hat{S}_{qi} = \sum_x \hat{S}_{qx} \delta_{xi} \ ,$$

where δ is a delta function. This new operator leads to

$$
\begin{aligned}
\rho^S(t) &= \sum_{qi} \lambda_i \left(\sum_x \hat{S}_{qx}\delta_{xi}\right) \Pi_i^S \left(\sum_y \hat{S}_{qy}^{\dagger}\delta_{yi}\right) \\
&= \sum_{qi} \lambda_i \left(\sum_x \hat{S}_{qx}\delta_{xi}\right) \Pi_i^S \Pi_i^S \Pi_i^S \left(\sum_y \hat{S}_{qy}^{\dagger}\delta_{yi}\right) \ ,
\end{aligned}
$$

where $\Pi_i \Pi_i \Pi_i = \Pi_i^2 \Pi_i = \Pi_i^2 = \Pi_i$. The orthogonality of projectors leads to $\Pi_i \Pi_x = \delta_{xi} \Pi_i$ or

$$\hat{S}_{qx}\delta_{xi} \Pi_i^S = \hat{S}_{qx} \Pi_x^S \Pi_i^S \ .$$

[2]It is interesting to note that almost all quantum states will have a non-vanishing discord [34]. The zero discord states discussed here are a special form of the initial composite state that leads to completely positive reduced dynamics, and that fact is the only reason they are of interest to this discussion.

This relation leads to

$$
\begin{aligned}
\rho^S(t) &= \sum_{qi} \lambda_i \left(\sum_x \hat{S}_{qx} \delta_{xi} \Pi_i^S \right) \Pi_i^S \left(\sum_y \Pi_i^S \delta_{yi} \hat{S}_{qy}^\dagger \right) \\
&= \sum_{qi} \lambda_i \left(\sum_x \hat{S}_{qx} \Pi_x^S \Pi_i^S \right) \Pi_i^S \left(\sum_y \Pi_i^S \Pi_y^S \hat{S}_{qy}^\dagger \right) \\
&= \sum_{qi} \lambda_i \left(\sum_x \hat{S}_{qx} \Pi_x^S \right) \Pi_i^S \Pi_i^S \Pi_i^S \left(\sum_y \Pi_y^S \hat{S}_{qy}^\dagger \right) \\
&= \sum_{qi} \lambda_i \left(\sum_x \hat{S}_{qx} \Pi_x^S \right) \Pi_i^S \left(\sum_y \Pi_y^S \hat{S}_{qy}^\dagger \right) \\
&= \sum_{qi} \lambda_i E_q \Pi_i^S E_q^\dagger \\
&= \sum_q E_q \rho^S E_q^\dagger \ ,
\end{aligned}
$$

where

$$
\rho^S = \left(\left(\rho^S \right)^\# \right)^\flat = \left(\sum_i \lambda_i \Pi_i^S \otimes \rho_i^B \right)^\flat = \sum_i \lambda_i \Pi_i^S \ , \tag{6.4}
$$

and $E_q \equiv \sum_m \hat{S}_{qm} \Pi_m^S$. The basic property of conjugate transposes, $(AB)^\dagger = B^\dagger A^\dagger$, was used in conjunction with the Hermiticity of projectors, i.e., $\Pi = \Pi^\dagger$, in the above derivation. Notice,

$$
\sum_q E_q^\dagger E_q = \sum_{qm} \Pi_m^S \hat{S}_{qm}^\dagger \hat{S}_{qm} \Pi_m^S = \sum_m \Pi_m^S = I \ ,
$$

where $\sum_q \hat{S}_{qm}^\dagger \hat{S}_{qm} = I$ was explained in the introduction of the $\hat{S}_{qi}$ operators (i.e., Eq. 4.3). Hence, by the representation theorem of quantum operations, zero discord initial composite states lead to completely positive reduced dynamics. Notice that initial composite states with no correlation (i.e., product states) are forms of zero discord states. This result was first shown in [78].

The discord of a quantum state is basis dependent in the sense that it depends on the specific projectors Π_i^S used to define the discord. The proof that zero discord states always lead to completely positive reduced dynamics is similarly basis dependent. The reduced dynamics are always completely positive for any composite dynamics if the reduced system can be written in the form of Eq. 6.4; i.e., the complete positivity of the reduced dynamics in this proof require that the reduced state system be a convex sum of a given complete set of projectors $\{\Pi_i\}$. In particular, this limitation means that the zero discord result presented here is not very useful for channels defined with tomography vectors.

To see this point, notice that a tomography vector $\vec{\tau}^\#$ might consist of composite states that have zero discord with respect to different projectors, but the reduced dynamics associated to $\vec{\tau}^\#$

will only be completely positive if every reduced state in $\vec{\tau}$ could be written in the zero discord form of Eq. 6.4 with respect to the same set of projectors $\{\Pi_i\}$. However, if every reduced state in $\vec{\tau}$ could be written in this way, then $\vec{\tau}$ would not be a tomography vector (because it would not be a basis for the reduced system). For example, consider the states of the canonical qubit tomography vector. Neither $|+\rangle\langle+|$ nor $|+_i\rangle\langle+_i|$ can be written in a zero discord form using $\{|0\rangle\langle0|, |1\rangle\langle1|\}$, and similar troubles arise trying to use $\{|+\rangle\langle+|, |-\rangle\langle-|\}$ or $\{|+_i\rangle\langle+_i|, |-_i\rangle\langle-_i|\}$ as the projector sets for a zero discord form of the states in $\vec{\tau}$. For this reason, we will not be using the zero discord result for anything beyond noting the result of this section.

It should be noted that a recent result in the literature (which has turned out to be very popular) makes the claim that a vanishing quantum discord is necessary and sufficient for a map to be completely positive given any composite dynamics [87]. This result was disputed within the community [19, 81], and the authors of [87] have recently declared the problem to still be open [30].

6.4   COMPLETELY POSITIVE REDUCED DYNAMICS WITHOUT A SPECIAL FORM

The results of the previous subsections spell out situations in which completely positive reduced dynamics are guaranteed.

- Regardless of initial correlations (i.e., for any initial composite state), local unitary composite evolution always lead to completely positive reduced dynamics.

- Regardless of coupling (i.e., for any composite system evolution), initial composite product states always lead to completely positive reduced dynamics.

These two situations are not the only situations in which completely positive reduced dynamics arise. Examples of completely positive and non-completely positive channels that do not fit into either one of these categories will be discussed in detail in later sections. But, now we will consider more general forms of the reduced dynamics that lead to complete positivity, but that involve conditions on both the correlation and the coupling.

Consider that, by definition, the reduced state of the system is the state of the composite system with the bath "traced out," i.e.,

$$\rho_f^S = \left(\rho_f^{SB}\right)^\flat \ ,$$

with $\rho^{SB} \in \mathcal{S}(\mathcal{H}^{SB})$ and $\rho^S \in \mathcal{S}(\mathcal{H}^S)$ and the subscripts mean "final" states. The dynamics of the composite system are defined by the composite unitary evolution and some initial reduced system state ρ^S as

$$\rho_f^{SB} = U^{SB}\left(\rho^S\right)^\sharp \left(U^{SB}\right)^\dagger \ .$$

Making the direct substitution yields,

$$\rho_f^S = \left(U^{SB} \left(\rho^S \right)^{\#} \left(U^{SB} \right)^{\dagger} \right)^{\flat} . \tag{6.5}$$

Expand the operators as

$$\rho^{SB} = \sum_i \lambda_i x_i^S \otimes y_i^B \tag{6.6}$$

and

$$U^{SB} = \sum_j v_j S_j \otimes B_j , \tag{6.7}$$

where $\rho^{SB} \in \mathcal{S}(\mathcal{H}^{SB})$ and $U^{SB} \in \mathcal{B}(\mathcal{H}^{SB})$ exist in the composite space, $x_i^S, S_j \in \mathcal{B}(\mathcal{H}^S)$ are restricted to the reduced system space, $y_i^B, B_j \in \mathcal{B}(\in \mathcal{H}^B)$ are restricted to the bath space, and $\lambda_i, v_j \in \mathbb{C}$. Notice that x_i^S and y_i^B are not necessarily valid density matrices and S_j and B_j are not necessarily unitary. This notation can be directly linked to the notation of Sec. 4.2 by noticing that the above constants are actually products of the constants in that section and the indices i and j above are actually single index labels for the multi-index sums of that section. The operators in the above sums are not necessarily unique, orthonormal, or a basis. The above notation is just a convenient shorthand for the more cumbersome notation of Sec. 4.2. In this new notation, the derivation becomes

$$\rho_f^S = \left(\left(\sum_j v_j S_j \otimes B_j \right) \left(\sum_i \lambda_i x_i^S \otimes y_i^B \right) \left(\sum_k v_k^* S_k^{\dagger} \otimes B_k^{\dagger} \right) \right)^{\flat} \tag{6.8}$$

$$= \sum_{ijk} \lambda_i v_j v_k^* \operatorname{Tr} \left(B_j y_i^B B_k^{\dagger} \right) S_j x_i^S S_k^{\dagger} \tag{6.9}$$

$$= \sum_{ijk} \lambda_i \alpha_{ijk} \bar{S}_j x_i^S \bar{S}_k^{\dagger} , \tag{6.10}$$

with $\alpha_{ijk} = \operatorname{Tr} \left(B_j y_i^B B_k^{\dagger} \right)$, $\bar{S}_j = v_j S_j$, and $\bar{S}_k^{\dagger} = v_k^* S_k^{\dagger}$.

This description of the reduced system dynamics is always true, but it is not always completely positive. By the representation theorem, if these reduced dynamics were completely positive, then they should have a Kraus operator sum form. The question of complete positivity is

$$\sum_x E_x \rho^S E_x^{\dagger} \stackrel{?}{=} \sum_{ijk} \lambda_i \alpha_{ijk} \bar{S}_j x_i^S \bar{S}_k^{\dagger} , \tag{6.11}$$

where $E_x \in \mathcal{B}(\mathcal{H}^S)$ is some arbitrary operator on the reduced system obeying the condition $\sum_x E_x^{\dagger} E_x = I$, and the initial state of the reduced system is

$$\rho^S = \left(\left(\rho^S \right)^{\#} \right)^{\flat} . \tag{6.12}$$

The reduced dynamics are completely positive when $\exists\ E_x$ such that the equality of Eq. 6.11 is satisfied. So, attempting to find such an E_x can be seen as a tool for determining the complete positivity of the reduced dynamics.

We will consider the results that have already been discussed to show some quick examples of trying to rewrite the general form of the reduced dynamics in a Kraus form to determine complete positivity. If $y_i^B = \rho^B = |\phi_f\rangle\langle\phi_f|\ \forall\ i$ with $\rho^B \in \mathcal{S}(\mathcal{H}^B)$, then

$$\alpha_{ijk} = \alpha_{jk} = \mathrm{Tr}\left(B_j \rho^B B_k^\dagger\right) = \sum_x \langle\phi_x| B_j |\phi_f\rangle\langle\phi_f| B_k^\dagger |\phi_x\rangle$$

where $\{|\phi_x\rangle\}$ is an orthonormal basis of $\mathcal{H}^B$ and $|\phi_f\rangle$ is some fixed state of the bath. In this case, Eq. 6.11 is satisfied with

$$E_x = \sum_j \sqrt{\langle\phi_x| B_j |\phi_f\rangle}\, \bar{S}_j$$

and

$$E_x^\dagger = \sum_k \sqrt{\langle\phi_x| B_k |\phi_f\rangle}\, \bar{S}_k \ .$$

This result is just the already proven statement that the reduced dynamics are completely positive, independent of the composite dynamics, if the bath is in some fixed pure state. This case falls under the second bullet point at the beginning of this section.

If the composite dynamics can be written in local unitary form, then $U^{SB} = U^S \otimes U^B$ with $U^S \in \mathcal{B}(\mathcal{H}^S)$ and $U^B \in \mathcal{B}(\mathcal{H}^B)$, both of which must be unitary because U^{SB} is unitary. The bath modifier term in the reduced dynamics then becomes

$$\alpha_{ijk} = \alpha_i = \mathrm{Tr}\left(U^B y_i^B (U^B)^\dagger\right) = \mathrm{Tr}\left((U^B)^\dagger U^B y_i^B\right) = \mathrm{Tr}\left(y_i^B\right)$$

by the cyclic property of the trace. The initial reduced system state would be

$$\rho^S = \left(\rho^{SB}(t)\right)^b = \left(\sum_i \lambda_i x_i^S \otimes y_i^B\right)^b = \sum_i \lambda_i \alpha_i x_i^S \ ,$$

and Eq. 6.11 would be satisfied with $E_x = \sum_j \bar{S}_j$ and $E_x^\dagger = \sum_k \bar{S}_k^\dagger$ where there is only one term in the sum over x. This result is the statement that local unitary composite dynamics lead to complete positivity for any initial composite state. This case is the first bullet point at the beginning of this section.

In a more general case, Eq. 6.11 will be satisfied if $\alpha_{ijk} = \kappa$, where $\kappa \geq 0$ is some constant, with $E_x = \sqrt{\kappa} \sum_j \bar{S}_j$ and $E_x^\dagger = \sqrt{\kappa} \sum_k \bar{S}_k^\dagger$.

Example 6.2   Consider again the qubit channel with a qubit bath and define the composite dynamics as[3]

$$U^{SB} = \frac{1}{2} \left(\sigma_0 \otimes \sigma_0 + \sigma_0 \otimes \sigma_1 - \sigma_3 \otimes \sigma_1 + \sigma_3 \otimes \sigma_0 \right) \ .$$

If the initial composite state is

$$\rho^{SB} = \rho \otimes \left(R\rho R^\dagger \right)$$

where ρ is some valid density matrix and

$$R = \begin{pmatrix} \cos\phi & -\sin\phi \\ \sin\phi & \cos\phi \end{pmatrix} ,$$

then the reduced dynamics become

$$
\begin{aligned}
\rho_f^S \ &= \ \frac{1}{4} \left((\sigma_0 \otimes \sigma_0 + \sigma_0 \otimes \sigma_1 - \sigma_3 \otimes \sigma_1 + \sigma_3 \otimes \sigma_0) \right. \\
&\qquad \left. \left(\rho \otimes \left(R\rho R^\dagger \right) \right) (\sigma_0 \otimes \sigma_0 + \sigma_0 \otimes \sigma_1 - \sigma_3 \otimes \sigma_1 + \sigma_3 \otimes \sigma_0) \right) \\
&= \ \frac{1}{4} \left(\alpha_{00} \left(\sigma_0 \rho \sigma_0 + \sigma_0 \rho \sigma_3 + \sigma_3 \rho \sigma_0 + \sigma_3 \rho \sigma_3 \right) \right. \\
&\qquad + \alpha_{01} \left(\sigma_0 \rho \sigma_0 - \sigma_0 \rho \sigma_3 + \sigma_3 \rho \sigma_0 - \sigma_3 \rho \sigma_3 \right) \\
&\qquad + \alpha_{10} \left(\sigma_0 \rho \sigma_0 + \sigma_0 \rho \sigma_3 - \sigma_3 \rho \sigma_0 - \sigma_3 \rho \sigma_3 \right) \\
&\qquad \left. + \alpha_{11} \left(\sigma_0 \rho \sigma_0 - \sigma_0 \rho \sigma_3 - \sigma_3 \rho \sigma_0 + \sigma_3 \rho \sigma_3 \right) \right) \ ,
\end{aligned}
$$

where the bath modifier terms are

$$
\begin{aligned}
\alpha_{00} \ &= \ \mathrm{Tr} \left(\sigma_0 R\rho R^\dagger \sigma_0 \right) = 1 \ , \\
\alpha_{11} \ &= \ \mathrm{Tr} \left(\sigma_1 R\rho R^\dagger \sigma_1 \right) = 1 \ , \\
\alpha_{01} \ &= \ \mathrm{Tr} \left(\sigma_0 R\rho R^\dagger \sigma_1 \right) = \mathrm{Tr} \left(R^\dagger \sigma_1 R\rho \right)
\end{aligned}
$$

and

$$\alpha_{10} = \mathrm{Tr} \left(\sigma_1 R\rho R^\dagger \sigma_0 \right) = \mathrm{Tr} \left(R^\dagger \sigma_1 R\rho \right) = \alpha_{01}$$

by the unitarity of R, the cyclic property of the trace, the unit trace condition of ρ, and the nice property of the Pauli matrices that $\sigma_m \sigma_n = \delta_{mn} \sigma_0 + \delta_{(m+n)1} \sigma_1$ when $m, n = 0, 1$. Plugging this back into the reduced dynamics yields

$$
\begin{aligned}
\rho_f^S \ &= \ \frac{1}{4} \left((\sigma_0 \rho \sigma_0 + \sigma_0 \rho \sigma_3 + \sigma_3 \rho \sigma_0 + \sigma_3 \rho \sigma_3) \right. \\
&\qquad + \alpha_{01} \left(\sigma_0 \rho \sigma_0 - \sigma_0 \rho \sigma_3 + \sigma_3 \rho \sigma_0 - \sigma_3 \rho \sigma_3 \right) \\
&\qquad + \alpha_{01} \left(\sigma_0 \rho \sigma_0 + \sigma_0 \rho \sigma_3 - \sigma_3 \rho \sigma_0 - \sigma_3 \rho \sigma_3 \right) \\
&\qquad \left. + \left(\sigma_0 \rho \sigma_0 - \sigma_0 \rho \sigma_3 - \sigma_3 \rho \sigma_0 + \sigma_3 \rho \sigma_3 \right) \right) \\
&= \ \frac{1}{4} \left(2 \left(\sigma_0 \rho \sigma_0 + \sigma_3 \rho \sigma_3 \right) + 2\alpha_{01} \left(\sigma_0 \rho \sigma_0 - \sigma_3 \rho \sigma_3 \right) \right) \ .
\end{aligned}
$$

---

[3]These composite dynamics are the famous controlled NOT gate. This gate will be discussed in much more detail in later sections.

Notice

$$\sigma_0 \rho \sigma_0 \pm \sigma_3 \rho \sigma_3 = (\sigma_0 \pm \sigma_3) \rho (\sigma_0 \pm \sigma_3) \mp \sigma_0 \rho \sigma_3 \mp \sigma_3 \rho \sigma_1 \ ,$$

which implies

$$
\begin{aligned}
\rho_f^S &= \frac{1}{2} \left(((\sigma_0 + \sigma_3) \rho (\sigma_0 + \sigma_3) - \sigma_0 \rho \sigma_3 - \sigma_3 \rho \sigma_0) \right. \\
&\quad \left. + \alpha_{01} ((\sigma_0 - \sigma_3) \rho (\sigma_0 - \sigma_3) + \sigma_0 \rho \sigma_3 + \sigma_3 \rho \sigma_0) \right) \\
&= |0\rangle \langle 0| \rho |0\rangle \langle 0| + \alpha_{01} |1\rangle \langle 1| \rho |1\rangle \langle 1| - \frac{[\rho, \sigma_3]}{2} (\alpha_{01} - 1) \ ,
\end{aligned}
$$

where the commutator of two matrices A and B is $[A, B] = AB - BA$ and the computational basis projectors of $|0\rangle \langle 0|$ and $|1\rangle \langle 1|$ are used to simplify the notation. This form of the reduced dynamics makes it clear that the dynamics are completely positive if $\alpha_{01} = 1$, i.e., Eq. 6.11 will be satisfied with $E_1 = |0\rangle \langle 0|$ and $E_2 = |1\rangle \langle 1|$ if $\alpha_{01} = 1$.

Notice that the bath modifier term α_{01} contains the product

$$R^\dagger \sigma_1 R = \begin{pmatrix} \sin(2\phi) & \cos(2\phi) \\ \cos(2\phi) & -\sin(2\phi) \end{pmatrix} \ ,$$

which is not equal to σ_0 for any ϕ. Also notice that $R^\dagger \sigma_1 R$ is not trace-preserving. These observations help point out that the condition of $\alpha_{01} = 1$ is a condition on both R and the initial state ρ. For example, if ρ is the computational ground state, i.e., $\rho = (\sigma_0 + \sigma_3)/2$, and $\phi = \pi/4$, then $R^\dagger \sigma_1 R (\sigma_0 + \sigma_3)/2 = (\sigma_3 + \sigma_0)/2$ and $\alpha_{01} = 1$. It can be shown that these reduced dynamics are completely positive if $\phi = \pi/4$ using Choi's theorem, which will be introduced in the next section. It is interesting to notice, however, that such a conclusion is difficult to make using only Eq. 6.11 as the test for complete positivity.

This example illustrates the fact that if the composite dynamics or initial composite state are not in the previously discussed specific forms, then whatever form the initial composite state takes will put restrictions on the possible forms of the composite dynamics that guarantee completely positive reduced dynamics, and vice versa. These results rely on the Kraus representation theorem of completely positive maps, but the next section will introduce another way of determining the complete positivity of a channel.

6.5 CHOI'S MATRIX

Suppose an experiment is conducted and the map describing the dynamics is not completely positive. How can an experimenter understand the lack of complete positivity?

The first step is to quantify the degree to which the complete positivity requirement for a map is not met. Consider a matrix defined as

$$C = (I \otimes \epsilon) \sum_{ij} |i\rangle \langle j| \otimes |i\rangle \langle j| = \sum_{ij} |i\rangle \langle j| \otimes \epsilon (|i\rangle \langle j|)$$

for some reduced dynamics ϵ where $|i\rangle\langle j|$ is the matrix of all zero entries with a 1 in the ijth position. A proof due to Choi [27] states that the map ϵ is completely positive if and only if C is positive semi-definite.[4]

The block diagonal of Choi's matrix consists of ϵ acting on valid states, but the rest of Choi's matrix must be found by tomography. As such, C can be written using a constructor on some tomography vector $\vec{\tau}$, i.e.,

$$C = \mathbf{C} \odot \vec{\tau}(t) = \mathbf{C} \odot \epsilon\left(\vec{\tau}\right),$$

where the constructor notation was introduced in Sec. 1.5. The map ϵ was introduced in Sec. 1.4.4, and in that section it was pointed out that ϵ is linear and Hermitian preserving. These conditions on ϵ imply that C is Hermitian [27, 88].

The spectral decomposition of C is assured by the Hermiticity, hence,

$$\mathbf{C} \odot \epsilon(\vec{\tau}) = \sum_r \Lambda_r |\Psi_r\rangle\langle\Psi_r| \ ,$$

where Λ_r and $|\Psi_r\rangle$ are the eigenvalues and eigenvectors of C. It follows that

$$\epsilon \text{ is completely positive} \Leftrightarrow \Lambda_r \geq 0 \ \forall r \ .$$

Choi's matrix is a common method for testing the complete positivity of tomography data in an experiment, and it extends the concept of complete positivity beyond the operator representation theorem. Choi's matrix will be the foundation of a quantification of the lack of complete positivity of a channel. This quantity will be called the negativity.

Before discussion turns to the negativity, a nice relationship between Choi's matrix and the Kraus decomposition needs to be addressed. Given a superoperator representation of a channel as

$$S = \mathbf{S} \odot \epsilon(\vec{\tau}) \ ,$$

the action of the channel on an arbitrary state ρ_i is

$$\text{mat}\left(S \text{ col}\left(\rho_i\right)\right) = \rho_f \ .$$

These concepts were introduced in Sec. 2. Given an operator sum form representation of the channel as

$$\rho_f = \sum_j c_j A_j \rho_i A_j^\dagger \ ,$$

it can be shown "by simple index gymnastics" [41] that

$$S = \sum_j c_j A_j^* \otimes A_j \ ,$$

---

[4]Notice $|i\rangle\langle j| \otimes |i\rangle\langle j|$ looks like a maximally entangled state. This notation is used to describe Choi's matrix as the application of the map ϵ to half of a maximally entangled state. Leung [57] outlines a method of performing process tomography based on Choi's proof and a similar logic.

where $c_j \in \mathbb{C}$ and A_j^* is the complex conjugate of A_j. The definition of Choi's matrix yields, for example in the single qubit case,

$$C = \mathbf{C} \odot \epsilon(\vec{\tau}) = \sum_{ij} |i\rangle\langle j| \otimes \epsilon\left(|i\rangle\langle j|\right) = \left(\begin{array}{c|c} \epsilon(|0\rangle\langle 0|) & \epsilon(|0\rangle\langle 1|) \\ \hline \epsilon(|1\rangle\langle 0|) & \epsilon(|1\rangle\langle 1|) \end{array} \right) ,$$

which is a 4×4 matrix. The eigenvectors of C are vectors that can be stacked into matrices as

$$\mathrm{mat}(|\Psi_r\rangle) \ .$$

Notice that a superoperator S can be constructed from C as [41]

$$S = \sum_r \Lambda_r \, \mathrm{mat}(|\Psi_r\rangle)^* \otimes \mathrm{mat}(|\Psi_r\rangle) \ ,$$

from which the operator sum representation follows from the action of S on some arbitrary state ρ_i as

$$\begin{aligned}
\mathrm{mat}\left(S \, \mathrm{col}(\rho_i)\right) &= \mathrm{mat}\left(\left(\sum_r \Lambda_r \, \mathrm{mat}(|\Psi_r\rangle)^* \otimes \mathrm{mat}(|\Psi_r\rangle)\right) \mathrm{col}(\rho_i)\right) \\
&= \sum_r \Lambda_r \, \mathrm{mat}(|\Psi_r\rangle) \rho_i \, \mathrm{mat}(|\Psi_r\rangle)^\dagger \ .
\end{aligned}$$

Choi's matrix is constructed from some tomography experiment characterized by a tomography vector $\vec{\tau}$ and is related to the superoperator representation of the channel through a re-stacking procedure. In fact, the Choi representation of a channel was originally introduced by Sudarshan and Jordan in 1961 [94] as a re-stacking of the superoperator representation. Choi's work in 1975 [27] cemented the relationship between the Choi matrix representation of a channel and complete positivity, but it was not the first introduction of the idea despite the naming conventions for these matrices.

Define a restacking procedure $(\cdot)^R$ such that

$$\begin{aligned}
C^R &= \left(\mathbf{C} \odot \epsilon(\vec{\tau})\right)^R \\
&= \left(\sum_{ij}^n |i\rangle\langle j| \otimes \epsilon\left(|i\rangle\langle j|\right)\right)^R \\
&= S \\
&= \left(\mathrm{col}\left(\epsilon\left(|0\rangle\langle 0|\right)\right) \cdots \mathrm{col}\left(\epsilon\left(|n\rangle\langle n|\right)\right)\right)
\end{aligned}$$

for some Choi matrix C and superoperator representation S of a channel characterized by a tomography vector $\vec{\tau}$ and the reduced dynamics ϵ.

Example 6.3   Given a single qubit Choi matrix

$$T = \begin{pmatrix} a & b & c & d \\ b^* & e & f & g \\ c^* & f^* & h & j \\ d^* & g^* & j^* & k \end{pmatrix}$$

the superoperator is found as

$$T^R = \begin{pmatrix} a & c^* & c & h \\ b^* & d^* & f & j^* \\ b & f^* & d & j \\ e & g^* & g & k \end{pmatrix}.$$

Such an operation is an involution, i.e.,

$$C = S^R = (C^R)^R ,$$

which can be seen clearly in the above example.

The Choi representation of a channel can be used to find the output state similar to the way the superoperator representation might be used, i.e., [46]

$$\text{mat}\left(S \, \text{col}(\rho_i)\right) = \rho_f = \text{Tr}_S\left(C\left(\rho_i^T \otimes I\right)\right)$$

where S and C are superoperator and Choi representations (respectively) of the same channel, $\rho_i, \rho_f \in \mathcal{S}(\mathcal{H}^S)$ are the initial and final states of the reduced system, I is the identity operator of the same dimension as ρ_i, Tr_S is the partial trace over the reduced system, and ρ^T is the transpose of ρ.

Any channel that can be tomographically characterized will have a Choi matrix representation, hence any such channel will have a operator sum form and superoperator representation defined in terms of the eigenvectors of the Choi matrix. Notice that if the channel is completely positive and has a Choi matrix $C = \sum_r \Lambda_r |\Psi_r\rangle\langle\Psi_r|$, then Kraus operators can be defined as

$$A_r = \sqrt{\Lambda_r} \, \text{mat}(|\Psi_r\rangle) ,$$

and the action of the channel on an arbitrary state ρ_i would have the Kraus sum form of

$$\rho_f = \sum_r A_r \rho_i A_r^\dagger ,$$

with $\sum_r A_r^\dagger A_r = \sum_r \Lambda_r \, \text{mat}(|\Psi_r\rangle))^\dagger \, \text{mat}(|\Psi_r\rangle) = I$ [41].

When the above form is derived from Choi's matrix, it is called the "canonical Kraus representation" of the channel. The eigendecomposition of Choi's matrix is not unique, from which follows the non-uniqueness of the Kraus representation.

Any channel ϵ with a Choi representation C can be written in an operator sum form as

$$\epsilon\left(\rho\right) = \sum_r \Lambda_r \left(\text{mat}\left(\left|\Psi_r\right\rangle\right)\right) \rho \left(\text{mat}\left(\left|\Psi_r\right\rangle\right)\right)^\dagger = \sum_r \Lambda_r B_r \rho B_r^\dagger \ ,$$

where $\{\Lambda_r\}$ are eigenvalues of C, $\{|\Psi_r\rangle\}$ are eigenvectors of C, $B_r = \text{mat}\left(|\Psi_r\rangle\right)$, and ρ is some arbitrary reduced system initial state. If $\Lambda_r \geq 0$ for all r, then this sum reduces to the Kraus representation as shown above. If, however, $\Lambda_r < 0$ for some r, then the operator sum can be written as

$$\epsilon\left(\rho\right) = \left(\sum_{r \in R} \Lambda_r B_r \rho B_r^\dagger\right) - \left(\sum_{k \in K} |\Lambda_k| B_k \rho B_k^\dagger\right) \equiv \left(\sum_{r \in R} B_r' \rho B_r'^\dagger\right) - \left(\sum_{k \in K} B_k' \rho B_k'^\dagger\right) \ ,$$

where $R = \{r | \Lambda_r \geq 0\}$, $K = \{k | \Lambda_k < 0\}$,

$$B_r' \equiv \sqrt{\Lambda_r} B_r$$

and

$$B_k' \equiv \sqrt{|\Lambda_k|} B_k \ .$$

The negative sign can be factored out in the case of a negative Choi representation, and in this way, any channel with a Choi representation can be written as the difference of two completely positive maps. Notice that both operator sums in the difference representation of ϵ are of the Kraus form (and hence completely positive), but they are not valid channels by themselves because they do not obey the trace preservation condition. To see this point, notice that, by definition, a channel must preserve the trace of the density matrix; hence,

$$\text{Tr}\left(\epsilon\left(\rho\right)\right) = 1 \ ,$$

but notice,

$$
\begin{aligned}
\text{Tr}\left(\epsilon\left(\rho\right)\right) &= \text{Tr}\left(\left(\sum_{r \in R} B_r' \rho B_r'^\dagger\right) - \left(\sum_{k \in K} B_k' \rho B_k'^\dagger\right)\right) \\
&= \text{Tr}\left(\sum_{r \in R} B_r' \rho B_r'^\dagger\right) - \text{Tr}\left(\sum_{k \in K} B_k' \rho B_k'^\dagger\right)
\end{aligned}
$$

by the linearity of the trace. If each term in the difference form was trace preserving, then

$$\text{Tr}\left(\sum_{r \in R} B_r' \rho B_r'^\dagger\right) = 1 = \text{Tr}\left(\sum_{k \in K} B_k' \rho B_k'^\dagger\right)$$

and

$$\text{Tr}\left(\epsilon\left(\rho\right)\right) = 0 \ ,$$

which contradicts the definition of ϵ as trace preserving. Therefore, the terms in the difference form cannot preserve the trace of the density matrix and cannot be treated as valid channels, even though they are both linear, completely positive maps.

Choi's theorem reduces the question of complete positivity for a channel to the question of positive semi-definiteness of the Choi matrix. This question is not easy, but it is straightforward. As such, the Choi matrix representation of a channel will be our preferred representation, although both the superoperator and operator sum representations will be used when it is convenient.

6.6 NEGATIVITY

The lack of complete positivity of a quantum channel will be called the negativity of the map ϵ, denoted as η_ϵ. This quantity will be defined in terms of the total sum of the eigenvalues and the sum of the negative eigenvalues of the Choi matrix representation of the channel as follows:

$$\eta_\epsilon = \frac{\sum_{k \in K} |\Lambda_k|}{\sum_r |\Lambda_r|} \, , \qquad (6.13)$$

where $K = \{k | \Lambda_k < 0\}$ and the Choi matrix has a spectral decomposition $C = \sum_r \Lambda_r |\Psi_r\rangle\langle\Psi_r|$. If $\mathcal{C}$ is the set of all Choi representations of quantum channels, then it will be shown later that $\eta_\epsilon : \mathcal{C} \to [0, 1/2)$.

The denominator is the sum of the absolute values of all of the eigenvalues of C and the numerator is the sum of the absolute value of all the negative eigenvalues of C. Notice

$$C \geq 0 \Rightarrow \sum_i |\Lambda_i| = 0 \Rightarrow \eta_\epsilon = 0 \, ,$$

and if every eigenvalue were negative, then

$$\sum_i |\Lambda_i| = \sum_r |\Lambda_r| \Rightarrow \eta_\epsilon = 1 \, ;$$

however, the trace condition of C also implies

$$\text{Tr}(C) > 0 \Rightarrow \max(\eta_\epsilon) < 1 \, ;$$

at least one eigenvalue of C will be non-negative because the trace of C is positive.

Hence, $\eta_\epsilon = 0 \Leftrightarrow \epsilon$ is completely positive and $\eta_\epsilon > 0 \Leftrightarrow \epsilon$ is negative. Also notice

$$\sum_r |\Lambda_r| \geq \text{Tr}(C),$$

which implies the negativity is never undefined.

Notice that the trace norm of a Hermitian matrix M can be defined as the absolute sum of its eigenvalues [102], i.e.,

$$||M||_1 \equiv \sum_i |\lambda_i| \ ,$$

where $||M||_1$ is the trace norm of M and λ_i is the ith eigenvalue of M. The trace norm of the Choi representation of a channel is precisely the denominator in the definition of the channel negativity. As such, given a channel with some Choi representation C, the negativity can be written as

$$\eta = \frac{\sum_i |\Lambda_i|}{||C||_1} \ ,$$

where Λ_i is the ith negative eigenvalue of C. Notice

$$\mathrm{Tr}(C) = \sum_{j \in R} |\Lambda_j| - \sum_{i \in K} |\Lambda_i| \ ,$$

where $R = \{r | \Lambda_r \geq 0\}$, $K = \{k | \Lambda_k < 0\}$, and Λ is, as given above, an eigenvalue of C. Notice

$$||C||_1 = \sum_{j \in R} |\Lambda_j| + \sum_{i \in K} |\Lambda_i| \ .$$

It follows that

$$\mathrm{Tr}(C) - ||C||_1 = -2 \sum_{i \in K} |\Lambda_i|$$

or

$$\sum_{i \in K} |\Lambda_i| = \frac{1}{2} (||C||_1 - \mathrm{Tr}(C)) \ .$$

So, the negativity η of the channel represented by C can be written as

$$\eta = \frac{||C||_1 - \mathrm{Tr}(C)}{2||C||_1} = \frac{1}{2} \left(1 - \frac{\mathrm{Tr}(C)}{||C||_1} \right) \ .$$

The trace norm is lower bounded by the trace as

$$||C||_1 \geq \mathrm{Tr}(C)$$

which immediately implies $\max(\eta) = 1/2$, or

$$\eta \in \left[0, \frac{1}{2} \right) \ .$$

Notice that $\mathrm{Tr}(C) = 0$ does not necessarily imply $||C||_1 = 0$, so $\eta = 1/2$ is achievable according to the above expression. However, it has already been pointed at that $\mathrm{Tr}(C) \neq 0$. This property of Choi's matrix is a direct consequence of the trace-preserving property of the reduced dynamics, and it implies that, as long as C is a valid Choi's matrix, the negativity will never actually achieve $1/2$. As will be seen in the next chapter, most of the negativities we discuss are for two qubit composite systems with $\text{Tr}(C) = 2$ that will have maximum values much less than $1/2$.

The negativity can be found empirically for a given map allowing the experimenter to quantify the complete positivity violation of his reduced dynamics. The analytical difficulty in the calculation of the negativity for the general case means some simple questions are difficult to answer.

The negativity is useful in understanding the role of experimental parameters on the complete positivity of a given experiment. It will allow the exploration of questions of why given experiments are negative. It will be useful in showing that some tomography experiments can yield negative superoperators even if the tomography is "perfect." These ideas are the topic of the next section.

CHAPTER 7

Negative Channels

Are negative channels experimentally realizable? Are negative channels simply a product of preparation mistakes or errors in tomography? And, perhaps most interestingly, do negative channels have any practical use in quantum information technologies? We will sketch out some answers to these questions.

7.1 DEFINITIONS

The discussion of negative channels needs to begin with a few definitions. The terms "channel" and "experiment" will be given formal definitions in an attempt prevent confusion. To that end, we have the following.

Definition 7.1   A *channel* is defined by the tuple $(\vec{\tau}, \sharp, U^{SB})$, i.e., a channel is defined by the tomography experiment used to characterize it (the tomography vector $\vec{\tau}$), its relationship to the bath (the $\sharp$ operation), and the composite evolution (U^{SB}).

The three items in the channel definition define the reduced dynamics (along with the flat operator). Every channel will have a *Choi representation* (also called the Choi's matrix for the channel, see Secs. 6.5), a *superoperator representation* (see Secs. 2 and 6.5), and an operator sum representation (derived independently of complete positivity considerations from the Choi representation, see Sec. 6.5) in addition to the originally introduced form of the reduced dynamics in Sec. 1.4.4. A channel is a mathematical representation of a physical process and is defined by the experiment used to characterize that process.

The collection of all channels will be split into two parts: completely positive and negative channels. The ideas behind these groupings have already been introduced and discussed. The negativity is derived from the Choi representation of the channel and was introduced in Sec. 6.6. Both definitions below will depend on the concept of negativity.

Definition 7.2   A *negative* channel is a channel with a non-zero negativity.

Definition 7.3   A *completely positive* channel is a channel with a negativity equal to zero.

Completely positive channels have a Kraus operator sum representation (see Sec. 4.1), which negative channels do not, and negative channels may have a non-trivial positivity domain (see Sec. 3), which completely positive channels do not. The main idea here is that the two classes are defined in terms of the negativity, which can be measured in the lab. Hence, the existence of both classes of channels can be verified experimentally.

The reduced dynamics are experimentally characterized using tomography. The definitions above make it clear that any discussion of channels (in any of its representations) should, in general, include discussion of the tomography vector. However, it will be useful to define an object more general than a channel. Removing the tomography vector from the channel definition leads to the following definition:

Definition 7.4   A *climate* will be defined by the pair $(\sharp, U^{SB})$.

A climate will, in general, lead to many different channels. In some cases (e.g., in the case of local unitary composite dynamics), a completely positive channel could be defined independently of the tomography vector used to characterize it. In such cases, a climate will be equivalent to a channel.

This language will make it easier to address the questions of negative channels. For example, given a sharp operation and composite evolution (i.e., given a climate), many different negative channels can be produced by simply performing tomography with different tomography vectors. In some sense, climates seem more fundamental in the design of negative channels. The choice of tomography sets is more or less arbitrary, usually subject to the experimenter's concerns about convenience and availability. The climate, however, is usually the entire point of interest for the experimenter. Before pursuing this idea any further, we will discuss the general form of sharp operations in this work.

7.2   SHARP OPERATIONS FOR NEGATIVE CHANNELS

The sharp operation is required to have three mathematical properties: linearity, consistency, and positivity on some domain of states. These properties were discussed in depth when the sharp operation was originally introduced, but they can be summed up concisely as follows. Linearity is required because quantum channels are tomographically characterized, consistency is required because any assumptions made about the bath must coincide with the empirical evidence available in the lab, and positivity on some domain of states is required to preserve the statistical interpretation of the density matrices measured in the lab. These requirements beg the question of the mathematical form of the sharp operation.

To explore this idea further, consider a single qubit reduced system with a single qubit bath and some tomography vector $\vec{\tau}$. A straightforward form of the sharp operation is

$$\vec{\tau}^{\sharp} = \vec{\tau} \otimes \rho^{B} \ , \tag{7.1}$$

where ρ^{B} is some fixed state of the bath. This form guarantees complete positivity for the channel by Pechukas' theorem (see Sec. 1.4.3) and must be avoided in a search for negative channels.

The mathematical requirements for the sharp operation are more limiting than they might appear at first glance. If the tomography vector $\vec{\tau}$ only contains pure states,[1] then consistency and positivity suggests a sharp operation of the form

$$\vec{\tau}^{\sharp} = \vec{\tau} \otimes \vec{b} \ , \tag{7.2}$$

where $\{\vec{b}\}_i \in \mathcal{B}(\mathcal{H}^B)$ is the ith state in the vector of states $\vec{b}$. The rigorous proof of this statement can be found in the proofs of Pechukas' theorem (see references in Sec. 1.4.3), but it can be motivated by making a few observations. First, tracing out one qubit from a pair of qubits in an entangled state will yield a mixed state. If mixed states are not used in tomography vectors, then sharp operations leading to entangled initial composite states are not consistent. Second, notice that the above form is consistent because $\vec{b}$ contains valid (specifically, unit trace) density matrices.

The problem with Eq. 7.2 is conceptual. The idea of a bath in a state completely unrelated to the reduced system state is difficult to accept. An experiment will begin with a preparation procedure that will prepare both the reduced system and the bath; as such, the initial state of the bath will always be related to the initial state of the reduced system through the preparation procedure. Admittedly, this relationship might be so complicated that $\vec{b}$ may appear unrelated to $\vec{\tau}$. A random guess for $\vec{b}$, however, would not be a justifiable assumption. The bath is defined by the experimenter's inability to access it, so any proposed form of the sharp operator for a given experiment will necessarily involve some "guesswork." It is important, however, that such "guesses" be motivated in some way.

At this point, the experimenter is going to need to make some educated guesses about his preparation procedure. For example, suppose the reduced system is a single spin with a bath of another single spin, both of which are perfectly isolated from the rest of the lattice structure in which they reside. If the reduced system is prepared by applying a large magnetic field to the entire sample, it might be reasonable to assume the reduced system and bath will be prepared identically, i.e.,

$$\vec{\tau}^{\sharp} = \vec{\tau} \otimes \vec{\tau} \ .$$

Perhaps preparation of the reduced system causes the bath to be prepared in some rotated state, i.e.,

$$\vec{\tau}^{\sharp} = \vec{\tau} \otimes \left(U : \vec{\tau} : U^{\dagger} \right) \ . \tag{7.3}$$

These sharp operations represent "reasonable" (or "justifiable") assumptions about the behavior of the preparation procedure on the bath. Such assumptions must be verified in the lab through a

---

[1]In general, it is not required that tomography vectors only contain pure states. However, most experimental realizations of tomography use only pure state tomography vectors [46, 68]. So, we will assume tomography vectors only contain pure states both as a way to relate closely to existing tomography experiments and to simplify the mathematical form of the sharp operation. It should be emphasized that the lack of entanglement in the general form of the sharp operator presented here is by assumption of a tomography vector that only contains pure states.

comparison of measured and theoretical representations of the channel, but they are a first step for the experimentalist trying to model his environment.

Equation 7.2 is linear by definition. It is also consistent and positive on some domain of states because the tensor product of two valid density matrices will always be another valid density matrix and $\vec{b}$ is a vector of valid density matrices. Equation 7.2 meets all three mathematical requirements demanded of it, and it can be justified as reasonable assumptions about the preparation of the reduced system. This form of the sharp operation (specifically Eq. 7.3) will be the form used in the examples of the next few sections.

CHAPTER 8

Negative Climates with Diagonal Composite Dynamics

In an effort to study a tractable case of negative climates, consider a sharp operation defined by Eq. 7.3 and composite dynamics defined as

$$U^{SB} = D = \text{diag}\,(D_1, D_2, D_3, D_4) = \begin{pmatrix} D_1 & 0 & 0 & 0 \\ 0 & D_2 & 0 & 0 \\ 0 & 0 & D_3 & 0 \\ 0 & 0 & 0 & D_4 \end{pmatrix}.$$

This climate describes a two qubit composite system undergoing diagonal unitary composite dynamics with a sharp operation that prepares the bath in some unitary rotation of the reduced system. The composite dynamics may not be in local unitary form, and the sharp operation does not lead to completely positive reduced dynamics in any obvious way.

The simplicity of this climate makes it amenable to analytical studies, but it is not without some physical significance. The appearance of this climate in the lab will be discussed later, but first consider the reduced dynamics:

$$\varepsilon\left(\vec{\tau}\right) = \left(D : \vec{\tau}^{\#} : D^{\dagger}\right)^{\flat} .$$

The composite dynamics yield

$$\{\rho_f^{\#}\}_{mn} = D_i \{\rho_i^{\#}\}_{mn} D_j^* ,$$

where $\rho_{f,i}^{\#} \in \mathcal{S}(\mathcal{H}^{SB})$ is some general composite state, the subscripts f and i denote the final and initial states, x^* is the complex conjugate of x, and $\{A\}_{mn}$ is the mnth element of the matrix A. Applying the flat operator yields

$$\varepsilon\left(\rho\right) = \begin{pmatrix} r & o \\ o^* & 1-r \end{pmatrix} , \tag{8.1}$$

where

$$r = \{\rho_i^{\#}\}_{11} + \{\rho_i^{\#}\}_{22}$$

and

$$o = D_1 D_3^* \{\rho_i^{\#}\}_{13} + D_2 D_4^* \{\rho_i^{\#}\}_{24} .$$

Equation 8.1 can be used to define the channel produced by this climate given the canonical tomography vector, i.e.,

$$\varepsilon\left(\vec{\tau}\right) = \left(\varepsilon\left(|0\rangle\langle0|\right), \varepsilon\left(|+\rangle\langle+|\right), \varepsilon\left(|+_i\rangle\langle+_i|\right), \varepsilon\left(|1\rangle\langle1|\right)\right)$$

$$= \left(\begin{pmatrix} r_1 & o_1 \\ o_1^* & 1-r_1 \end{pmatrix}, \begin{pmatrix} r_2 & o_2 \\ o_2^* & 1-r_2 \end{pmatrix}, \begin{pmatrix} r_3 & o_3 \\ o_3^* & 1-r_3 \end{pmatrix}, \begin{pmatrix} r_4 & o_4 \\ o_4^* & 1-r_4 \end{pmatrix}\right),$$

with (just as above) $r_i = \{\varrho^\#\}_{11} + \{\varrho^\#\}_{22}$ and $o = D_1 D_3^* \{\varrho^\#\}_{13} + D_2 D_4^* \{\varrho^\#\}_{24}$ with $\varrho = \{\vec{\tau}\}_i$.

Given the transformation matrix

$$\hat{R} = \frac{1}{2}\begin{pmatrix} 2 & i-1 & -(1+i) & 0 \\ 0 & 2 & 2 & 0 \\ 0 & -2i & 2i & 0 \\ 0 & i-1 & -(1+i) & 2 \end{pmatrix},$$

the canonical tomography vector can be written in the standard basis as

$$\vec{s} = \vec{\tau}\hat{R} = \left(\begin{pmatrix} 1 & 0 \\ 0 & 0 \end{pmatrix}, \begin{pmatrix} 0 & 0 \\ 1 & 0 \end{pmatrix}, \begin{pmatrix} 0 & 1 \\ 0 & 0 \end{pmatrix}, \begin{pmatrix} 0 & 0 \\ 0 & 1 \end{pmatrix}\right).$$

The channel can then be rewritten as

$$\varepsilon\left(\vec{\tau}\right)\hat{R} = \left(\begin{pmatrix} r_1 & o_1 \\ o_1^* & 1-r_1 \end{pmatrix}, \begin{pmatrix} c_1 & k_1 \\ t_1 & d_1 \end{pmatrix}, \begin{pmatrix} c_1^* & t_1^* \\ k_1^* & d_1^* \end{pmatrix}, \begin{pmatrix} r_4 & o_4 \\ o_4^* & 1-r_4 \end{pmatrix}\right),$$

where

$$c_1 = \left(-\frac{1}{2}+\frac{i}{2}\right)(r_1 - (1+i)r_2 - (1-i)r_3 + r_4)$$

$$k_1 = \left(-\frac{1}{2}+\frac{i}{2}\right)(o_1 - (1+i)o_2 - (1-i)o_3 + o_4)$$

$$t_1 = \left(-\frac{1}{2}+\frac{i}{2}\right)(o_1^* - (1+i)o_2^* - (1-i)o_3^* + o_4^*)$$

$$d_1 = \left(\frac{1}{2}-\frac{i}{2}\right)(r_1 - (1+i)r_2 - (1-i)r_3 + r_4) .$$

The Choi representation of this channel follows immediately:

$$\mathbf{C}\odot\varepsilon(\vec{\tau}) = \begin{pmatrix} r_1 & o_1 & c_1^* & t_1^* \\ o_1^* & 1-r_1 & k_1^* & d_1^* \\ c_1 & k_1 & r_4 & o_4 \\ t_1 & d_1 & o_4^* & 1-r_4 \end{pmatrix} .$$

The sharp operation for this climate takes the form

$$\vec{\tau}^\# = \vec{\tau} \otimes \left(U : \vec{\tau} : U^\dagger\right) .$$

Write the unitary U in the Pauli basis as

$$U = \vec{u} \cdot \vec{\sigma} = \begin{pmatrix} u_1 + u_4 & u_2 - iu_3 \\ u_2 + iu_3 & u_1 - u_4 \end{pmatrix} ,$$

where $\vec{u} = (u_1, u_2, u_3, u_4) \in \mathbb{C}$ is some vector of complex numbers and $\vec{\sigma}$ is the Pauli vector of states introduced in Sec. 2. The constants used above can be rewritten in terms of just elements of $\vec{u}$ and D. Notice

$$\begin{aligned}
\vec{r} &= (r_1, r_2, r_3, r_4) \\
&= ((u_2 + iu_3)(u_2^* - iu_3^*) + (u_1 + u_4)(u_1^* + u_4^*), \\
&\quad \frac{1}{2}\left((u_1 + u_2)(u_1^* + u_2^*) + (u_3 + iu_4)(u_3^* - iu_4^*)\right), \\
&\quad \frac{1}{2}\left((u_1 + u_3)(u_1^* + u_3^*) + (u_2 - iu_4)(u_2^* + iu_4^*)\right), \\
&\quad 0)
\end{aligned}$$

and

$$\begin{aligned}
\vec{o} &= (o_1, o_2, o_3, o_4) \\
&= (0, \\
&\quad \frac{D_1 D_3^*}{4}(u_1 + u_2 - iu_3 + u_4)\left(u_1^* + u_2^* + iu_3^* + u_4^*\right) \\
&\quad + \frac{D_2 D_4^*}{4}(u_1 + u_2 + iu_3 - u_4)\left(u_1^* + u_2^* - iu_3^* - u_4^*\right), \\
&\quad -\frac{D_1 D_3^*}{4}(u_1 + iu_2 + u_3 + u_4)\left(u_2^* + i(u_1^* + u_3^* + u_4^*)\right)) \\
&\quad -\frac{D_2 D_4^* i}{4}(u_1 - iu_2 + u_3 - u_4)\left(u_1^* + iu_2^* + u_3^* - u_4^*\right), \\
&\quad 0) \ .
\end{aligned}$$

The unitarity of U requires

$$(u_1 + u_4)(u_1^* + u_4^*) + (u_2 + iu_3)(u_2^* - iu_3^*) = 1 \ ;$$

hence, $r_1 = r_1^* = 1$. It has already been shown that $r_4 = o_1 = o_4 = 0$. The unitarity of U similarly requires $r_2 = r_3 = 1/2$. The simplest way to see this fact is to rewrite U as

$$U = \begin{pmatrix} \alpha & \beta \\ \gamma & \delta \end{pmatrix}$$

and notice that, in this new notation for U, we have

$$r_2 = \frac{1}{4}\left((\alpha + \beta)(\alpha^* + \beta^*) + (\gamma + \delta)(\gamma^* + \delta^*)\right)$$

and

$$r_3 = \frac{1}{4}\left((\alpha + i\beta)\left(\alpha^* - i\beta^*\right) + (\gamma + i\delta)\left(\gamma^* - i\delta^*\right)\right) \ .$$

The unitarity requirement of U is $UU^\dagger = U^\dagger U = I$ where I is the identity matrix. This requirement implies the following system of equations

$$\alpha\alpha^* + \beta\beta^* = \gamma\gamma^* + \delta\delta^* = \alpha\alpha^* + \gamma\gamma^* = \beta\beta^* + \delta\delta^* = 1$$

and

$$\alpha\gamma^* + \beta\delta^* = \gamma\alpha^* + \delta\beta^* = \beta\alpha^* + \delta\gamma^* = \alpha\beta^* + \gamma\delta^* = 0 \ .$$

This system of equations, in turn, can be used to show that $r_2 = r_3 = 1/2$ by direct substitution. Notice that the unitarity of U and the choice of the canonical tomography vector fixes $\vec{r} = (1, 1/2, 1/2, 0)$. From above, it can be seen that $c_1 = -d_1$ and $c_1 = 0$ if $\vec{r} = (1, 1/2, 1/2, 0)$.

The diagonal structure of D and the simple tensor product form of the sharp operation leads to a relatively simple Choi representation for this channel:

$$C_G = \mathbf{C} \odot \varepsilon(\vec{\tau}) = \begin{pmatrix} 1 & 0 & 0 & t_1^* \\ 0 & 0 & k_1^* & 0 \\ 0 & k_1 & 0 & 0 \\ t_1 & 0 & 0 & 1 \end{pmatrix} ,$$

where $\varepsilon(\vec{\tau})$ is a vector of states containing the process tomography information for the channel $(\vec{\tau}, \vec{\tau} \otimes (U : \vec{\tau} : U^\dagger), D)$ which is found using the canonical tomography vector $\vec{\tau}$. The last of the variables from this vast sea of notation can then be rewritten as follows:

$$
\begin{aligned}
k_1 &= \left(-\frac{1}{2} + \frac{i}{2}\right)(o_1 - (1+i)o_2 - (1-i)o_3 + o_4) \\
&= \left(\frac{1}{4} + \frac{i}{4}\right)(D_2 D_4^*(-i(u_1 - u_4)(u_2^* - iu_3^*) \\
&\quad + (u_2 + iu_3)(u_1^* - u_4^*)) + D_1 D_3^*((u_1 + u_4)(u_2^* + iu_3^*) - i(u_2 - iu_3)(u_1^* + u_4^*))) \\
t_1 &= \left(-\frac{1}{2} + \frac{i}{2}\right)(o_1^* - (1+i)o_2^* - (1-i)o_3^* + o_4^*) \\
&= \frac{1}{4}\left(D_4 D_2^*(u_1 + u_2 + iu_3 - u_4)(u_1^* + u_2^* - iu_3^* - u_4^*) \right. \\
&\quad + D_4 D_2^*(u_1 - iu_2 + u_3 - u_4)(u_1^* + iu_2^* + u_3^* - u_4^*) \\
&\quad + D_3 D_1^*(u_1 + iu_2 + u_3 + u_4)(u_1^* - iu_2^* + u_3^* + u_4^*) \\
&\quad \left. + D_3 D_1^*(u_1 + u_2 - iu_3 + u_4)(iu_3^* + (u_1 + u_2 + u_4)^*)\right) \ .
\end{aligned}
$$

Despite appearances to the contrary, the point of all of this algebra is not to simply confuse the reader. Consider a few different forms of the unitary rotation of the bath qubit. For example, suppose the bath qubit is not rotated at all:

$$\vec{u} = (1, 0, 0, 0) \Rightarrow \vec{r} = \left(1, \frac{1}{2}, \frac{1}{2}, 0\right), \vec{o} = \left(0, \frac{1}{4}\left(D_1 D_3^* + D_2 D_4^*\right), -\frac{i}{4}\left(D_1 D_3^* + D_2 D_4^*\right), 0\right) \ .$$

These vectors lead to $c_1 = d_1 = k_1 = 0$ and the Choi representation of this channel is

$$C_S = \mathbf{C} \odot \varepsilon'(\vec{\tau}) = \begin{pmatrix} 1 & 0 & 0 & t_1^* \\ 0 & 0 & 0 & 0 \\ 0 & 0 & 0 & 0 \\ t_1 & 0 & 0 & 1 \end{pmatrix} .$$

This example is simple enough that the spectrum of C_S can be written down directly as

$$\text{spec}(C_S) = \left(1 + \sqrt{t_1 t_1^*}, 1 - \sqrt{t_1 t_1^*}, 0, 0 \right) ,$$

and notice

$$
\begin{aligned}
t_1 t_1^* &= \frac{1}{4} \left(D_3 D_1^* + D_4 D_2^* \right) \left(D_3^* D_1 + D_4^* D_2 \right) \\
&= \frac{1}{4} \left(D_3 D_1^* D_3^* D_1 + D_4 D_2^* D_3^* D_1 + D_3 D_1^* D_4^* D_2 + D_4 D_2^* D_4^* D_2 \right) \\
&= \frac{1}{4} \left(2 + D_4 D_2^* D_3^* D_1 + D_3 D_1^* D_4^* D_2 \right) ,
\end{aligned}
$$

where the last equality follows from the unitarity of D, i.e., $D_j D_j^* = 1$. If this channel is completely positive, i.e., if it is known that $\eta_S = 0$, then

$$1 - \sqrt{t_1 t_1^*} \geq 0$$

and

$$1 + \sqrt{t_1 t_1^*} \geq 0 .$$

These expressions can be written more succinctly as[1]

$$1 \geq t_1 t_1^* \geq 0$$

which reduces to

$$2 \geq D_4 D_2^* D_3^* D_1 + D_3 D_1^* D_4^* D_2 \geq -2 .$$

The elements of D can be written in polar form as $D_j = \alpha_j e^{i\theta_j}$, which implies

$$D_4 D_2^* D_3^* D_1 = \alpha_1 \alpha_2 \alpha_3 \alpha_4 e^{i(\theta_1 + \theta_4 - \theta_2 - \theta_3)} = \alpha' e^{i\theta'} ,$$

with $\alpha' = \alpha_1 \alpha_2 \alpha_3 \alpha_4$ and $\theta' = \theta_1 + \theta_4 - \theta_2 - \theta_3$. The unitarity of D implies

$$D_j D_j^* = \alpha_j \alpha_j e^{i(\theta_j - \theta_j)} = 1 \Rightarrow \alpha_j = 1 \ \forall j \Rightarrow \alpha' = 1 .$$

---

[1] Notice $zz^* = (x + iy)(x - iy) = x^2 + y^2 \geq 0$ where $z = x + iy$ with $x, y \in \mathbb{R}$.

Notice

$$D_3 D_1^* D_4^* D_2 = \alpha' e^{i(\theta_3 + \theta_2 - \theta_1 - \theta_4)} = \alpha' e^{-i\theta'} \ ,$$

which implies

$$D_4 D_2^* D_3^* D_1 + D_3 D_1^* D_4^* D_2 = e^{i\theta'} + e^{-i\theta'} = 2\cos\theta' \ ,$$

where $\alpha' = 1$ as shown above and the last equality follows from Euler's formula [35]. The above inequality is reduced one last time to

$$1 \geq \cos\theta' \geq -1 \ ,$$

which is always true.

Hence, the channel defined by the canonical tomography vector, D, and the above sharp operation with U defined by $\vec{u} = (1, 0, 0, 0)$ has zero negativity. No restrictions were put on D other than unitarity. It is the sharp operation alone that guarantees the complete positivity of this channel. Interestingly, $\vec{u} = (0, 1, 0, 0)$, $\vec{u} = (0, 0, 1, 0)$, and $\vec{u} = (0, 0, 0, 1)$ all lead to the same vectors $\vec{r}$ and $\vec{o}$ as $\vec{u} = (1, 0, 0, 0)$. None of these sharp operations can led to negative channels.

Notice

$$k_1 = 0 \Rightarrow -(1+i)o_2 - (1-i)o_3 = 0 \Rightarrow o_2 = \frac{i-1}{i+1}o_3 \ .$$

This condition for the complete positivity of C_G is satisfied by $\vec{u} = (1, 0, 0, 0)$, $\vec{u} = (0, 1, 0, 0)$, $\vec{u} = (0, 0, 1, 0)$, and $\vec{u} = (0, 0, 0, 1)$, but it is not satisfied in general for C_G. This condition can be written in terms of $\vec{u}$ and D as

$$k_1 = 0 \Rightarrow \frac{n}{m} = \frac{1-i}{1+i} \ ,$$

with

$$
\begin{aligned}
n &= D_1 D_3^* (u_1 + u_2 - iu_3 + u_4)(u_1^* + u_2^* + iu_3^* + u_4^*) \\
&\quad + D_2 D_4^* (u_1 + u_2 + iu_3 - u_4)(u_1^* + u_2^* - iu_3^* - u_4^*) \\
m &= D_1 D_3^* (u_1 + iu_2 + u_3 + u_4)(u_2^* + i(u_1^* + u_3^* + u_4^*))) \\
&\quad + D_2 D_4^* i (u_1 - iu_2 + u_3 - u_4)(u_1^* + iu_2^* + u_3^* - u_4^*) \ .
\end{aligned}
$$

This complicated condition is met when C_G is completely positive, but the utility is limited by the complexity of the equation. It does, however, serve as a nice illustration of the dependence of the negativity on both the sharp operation (i.e., $\vec{u}$) and the composite dynamics (i.e., D).

This seemingly endless sidetrack of linear algebra is more useful than simply illustrating the role of initial conditions and composite dynamics in this climate. It also helps point the way to a sharp operation that might lead to negative dynamics. For example, consider $\vec{u} = (0, 2^{-1/2}, 0, 2^{-1/2})$ (i.e., $U = H_d$, the initial bath qubit state is a Hadamard rotation of the reduced system initial state):

$$\vec{u} = \left(0, \frac{1}{\sqrt{2}}, 0, \frac{1}{\sqrt{2}}\right) \Rightarrow \vec{r} = \left(1, \frac{1}{2}, \frac{1}{2}, 0\right), \vec{o} = \left(0, \frac{D_1 D_3^*}{2}, -\frac{i}{4}\left(D_1 D_3^* + D_2 D_4^*\right), 0\right) \ .$$

These vectors imply

$$k_1 = \left(\frac{i}{2} - \frac{1}{2}\right)\left(\frac{-(1+i)}{2}D_1 D_3^* + \frac{i(1-i)}{4}\left(D_1 D_3^* + D_2 D_4^*\right)\right) = \frac{1}{4}\left(D_1 D_3^* - D_2 D_4^*\right)$$

and

$$t_1 = \left(\frac{i}{2} - \frac{1}{2}\right)\left(\frac{-(1+i)}{2}D_1^* D_3 - \frac{i(1-i)}{4}\left(D_1^* D_3 + D_2^* D_4\right)\right) = \frac{1}{4}\left(3 D_1^* D_3 + D_2^* D_4\right) \quad .$$

This sharp operation will to lead to completely positive dynamics if k_1 vanishes, which will only happen if $D_1 D_3^* = D_2 D_4^*$. Hence, the sharp operation defined by $\vec{u} = (0, 2^{-1/2}, 0, 2^{-1/2})$ might led to a negative channel. Similarly,

$$\vec{u} = \left(0, \frac{i}{\sqrt{2}}, 0, \frac{i}{\sqrt{2}}\right) \Rightarrow \vec{r} = \left(1, \frac{1}{2}, \frac{1}{2}, 0\right), \vec{o} = \left(0, \frac{D_1 D_3^*}{2}, -\frac{i}{4}\left(D_1 D_3^* + D_2 D_4^*\right), 0\right)$$

can also lead to a negative channel.

These two specific examples are not the only examples. The conditions for this channel to be completely positive are written in several different ways above and they are quite restrictive.

As a quick aside, notice that the condition for the positivity of Eq. 8.1 is

$$1 \geq 1 + 4r(r - 1) + 4oo^* \Rightarrow r(1 - r) \geq oo^* \quad .$$

The abstract algebra can be difficult to parse and even harder to understand physically, so the next section will focus exclusively on the $U = H_d$ example of a negative channel arising from diagonal composite dynamics.

8.1 SPECIFIC NEGATIVE CHANNEL EXAMPLE: $\left(\vec{\tau}, \vec{\tau} \otimes \left(H_d : \vec{\tau} : H_d^\dagger\right), D\right)$

The channel described by the composite dynamics

$$D = \text{diag}(D_1, D_2, D_3, D_4) \quad ,$$

the sharp operation

$$\vec{\tau}^\# = \vec{\tau} \otimes \left(H_d : \vec{\tau} : H_d^\dagger\right) \quad ,$$

with

$$H_d = \frac{1}{\sqrt{2}}(\sigma_1 + \sigma_3) \quad ,$$

and the usual tomography vector

$$\vec{\tau} = (|0\rangle\langle 0|, |+\rangle\langle +|, |+_i\rangle\langle +_i|, |1\rangle\langle 1|) \quad ,$$

is not necessarily completely positive. The reduced dynamics are

$$
\begin{aligned}
\varepsilon\left(\vec{\tau}\right) &= \left(D : \vec{\tau}^{\#} : D^{\dagger}\right)^{\flat} \\
&= \left(\begin{pmatrix} 1 & 0 \\ 0 & 0 \end{pmatrix}, \frac{1}{2}\begin{pmatrix} 1 & D_1 D_3^* \\ D_3 D_1^* & 1 \end{pmatrix}, \right. \\
&\quad \left. \frac{1}{4}\begin{pmatrix} 2 & -i(D_1 D_3^* + D_2 D_4^*) \\ i(D_3 D_1^* + D_4 D_2^*) & 2 \end{pmatrix}, \begin{pmatrix} 0 & 0 \\ 0 & 1 \end{pmatrix}\right) .
\end{aligned}
$$

The transformation matrix $\hat{R}$ from the previous subsection yields

$$
\begin{aligned}
\varepsilon\left(\vec{\tau}\right)\hat{R} &= \left(\begin{pmatrix} 1 & 0 \\ 0 & 0 \end{pmatrix}, \right. \\
&\quad \frac{1}{4}\begin{pmatrix} 0 & D_1 D_3^* - D_2 D_4^* \\ 3D_3 D_1^* + D_4 D_2^* & 0 \end{pmatrix}, \\
&\quad \frac{1}{4}\begin{pmatrix} 0 & 3D_1 D_3^* + D_2 D_4^* \\ D_3 D_1^* - D_4 D_2^* & 0 \end{pmatrix}, \\
&\quad \left. \begin{pmatrix} 0 & 0 \\ 0 & 1 \end{pmatrix}\right) ,
\end{aligned}
$$

and the Choi representation of the channel immediately follows as

$$
C_H = \mathbf{C} \odot \varepsilon(\vec{\tau}) = \begin{pmatrix} 1 & 0 & 0 & x \\ 0 & 0 & y & 0 \\ 0 & y^* & 0 & 0 \\ x^* & 0 & 0 & 1 \end{pmatrix}
$$

with

$$
x = \frac{1}{4}\left(3D_3 D_1^* + D_4 D_2^*\right)
$$

and

$$
y = \frac{1}{4}\left(D_1 D_3^* - D_2 D_4^*\right) .
$$

The Choi representation of this channel has the following spectrum:

$$
\text{spec}(C_H) = \left(1 - \sqrt{xx^*}, 1 + \sqrt{xx^*}, -\sqrt{yy^*}, \sqrt{yy^*}\right) .
$$

Following the reasoning of the previous subsection, the elements of D can be written in polar form as $D_j = e^{i\theta_j}$, which leads to

$$
xx^* = \frac{1}{8}\left(5 + 3\cos\left(f_\theta\right)\right)
$$

and

$$yy^* = \frac{1}{8}\left(1 - \cos\left(f_\theta\right)\right) = \frac{1}{4}\sin^2\left(\frac{f_\theta}{2}\right) \; ,$$

with $f_\theta = \theta_1 - \theta_2 - \theta_3 + \theta_4$ and where the last equality uses the sine half angle formula (i.e., $2\sin^2\theta = 1 - \cos 2\theta$).

Notice that C_H represents a completely positive channel only when $yy^* = 0$, i.e., when $f_\theta = 2n\pi$ where $n \in \mathbb{Z}$ is some integer. If $f_\theta = 2n\pi$, then $xx^* = 1$, $C_H \geq 0$, and $\eta_H = 0$ (where η_H is the negativity of the channel represented by C_H).

The channel represented by C_H can be negative, and the negativity can be bounded as follows:

$$\cos(f_\theta) \in [-1, 1] \Rightarrow xx^* \in \left[\frac{1}{4}, 1\right] \Rightarrow 1 \pm \sqrt{xx^*} \in [0, 2]$$

and

$$\sin\left(\frac{f_\theta}{2}\right) \in [-1, 1] \Rightarrow yy^* \in \left[0, \frac{1}{4}\right] \Rightarrow \pm\sqrt{yy^*} \in \left[-\frac{1}{2}, \frac{1}{2}\right]$$

which implies

$$\max \eta_H = \frac{1}{6}$$

(which occurs at $f_\theta = n\pi$ with n odd). So,

$$\eta_H \in \left[0, \frac{1}{6}\right] \; .$$

An important property of this channel is that the composite dynamics D have a clear physical interpretation. The unitary evolution of the composite system can be written down as

$$U^{SB} = e^{-\frac{it}{\hbar}H^{SB}} \; ,$$

where H^{SB} is the time-independent Hamiltonian describing the composite dynamics and t is the elapsed time. The composite system Hamiltonian is a self-adjoint operator and will, therefore, be diagonalizable as

$$H^{SB} = V\Lambda V^{-1} \; ,$$

where V is a 4×4 matrix constructed from the eigenvectors of H^{SB} and $\Lambda = \mathrm{diag}(\lambda_1, \lambda_2, \lambda_3, \lambda_4)$ with λ_i being the ith eigenvalue of H^{SB}. The Hamiltonian operator H^{SB} is diagonal in its eigenbasis with diagonal entries corresponding to its eigenvalues, and in standard quantum mechanics this eigenbasis is called the "energy basis" or the "energy eigenbasis." The eigenvalues correspond to possible energy measurement values [56].

The matrix exponentiation of a diagonal matrix is straightforward, i.e.,

$$A = \mathrm{diag}(a_1, a_2, a_3, a_4) \Rightarrow e^A = \mathrm{diag}(e^{a_1}, e^{a_2}, e^{a_3}, e^{a_4}) \; .$$

Putting everything together: In the energy eigenbasis,

$$H^{SB} = \text{diag}(\lambda_1, \lambda_2, \lambda_3, \lambda_4)$$
$$\Rightarrow U^{SB} = \text{diag}\left(\exp\left\{-\frac{it\lambda_1}{\hbar}\right\}, \exp\left\{-\frac{it\lambda_2}{\hbar}\right\}, \exp\left\{-\frac{it\lambda_3}{\hbar}\right\}, \exp\left\{-\frac{it\lambda_4}{\hbar}\right\}\right) .$$

The diagonal composite dynamics D can then be interpreted as the energy eigenbasis evolution of the composite system governed by a time independent Hamiltonian. The elements of D have already been written in polar form as $D_j = e^{i\theta_j}$ which yields (by comparison with the above expressions)

$$\theta_j = -\frac{t\lambda_j}{\hbar}$$

and

$$f_\theta = -\frac{t}{\hbar}(\lambda_1 - \lambda_2 - \lambda_3 + \lambda_4) . \tag{8.2}$$

The condition of complete positivity for D,

$$f_\theta = 2n\pi \Rightarrow \eta_H = 0 ,$$

can now be interpreted in terms of the physical parameters of the Hamiltonian.

Example 8.1   Consider the toy Hamiltonian

$$H_{zz} = k_z(\sigma_3 \otimes \sigma_3) = \text{diag}(k_z, -k_z, -k_z, k_z) .$$

This Hamiltonian leads to

$$f_\theta = -\frac{4k_z t}{\hbar} ,$$

which has a vanishing negativity if

$$k_z = -\frac{\pi n\hbar}{2t} \tag{8.3}$$

(t is the elapsed time; $t = 0$ would imply no passage of time between two points of evolution and time is non-negative by convention, hence $t > 0$) and a maximum negativity if

$$k_z = -\frac{\pi n\hbar}{4t} .$$

The requirement of completely positive reduced dynamics would imply that the coupling constant k_z in H_{zz} has only very specific physically allowable values corresponding to Eq. 8.3. It may be that requiring completely positive reduced dynamics is unreasonable, but Eq. 8.3 does not imply such a strong statement. It does, however, point out that such a requirement leads to very strict constraints on allowable composite Hamiltonians.[2]

Consider the following examples.

Example 8.2   The Hamiltonian H_{zz} is already diagonal (and is, hence, already in a spectral representation), but the basis of H_{zz} can be changed by conjugation by a unitary matrix. For example,

$$V_1 = \begin{pmatrix} 0 & 0 & 0 & 1 \\ 0 & 1 & 0 & 0 \\ 1 & 0 & 0 & 0 \\ 0 & 0 & 1 & 0 \end{pmatrix} \tag{8.4}$$

leads to

$$H'_{zz} = V_1^{-1} H_{zz} V_1 = \mathrm{diag}(-k_z, -k_z, k_z, k_z) \ .$$

This rotated Hamiltonian yields

$$f_\theta = 0 \ ;$$

the un-rotated Hamiltonian H_{zz} can lead to a negative channel, but the rotated H'_{zz} is always completely positive. Similarly,

$$H''_{zz} = V_2^{-1} H_{zz} V_2 = \mathrm{diag}(k_z, k_z, -k_z, -k_z) \ ,$$

with

$$V_2 = \begin{pmatrix} 1 & 0 & 0 & 0 \\ 0 & 0 & 1 & 0 \\ 0 & 0 & 0 & 1 \\ 0 & 1 & 0 & 0 \end{pmatrix}$$

is, likewise, always completely positive. Notice $V_1 V_1^\dagger = V_2 V_2^\dagger = I$. The dynamics described by H_{zz} can be negative in the σ_z-basis (i.e., the basis of H_{zz}) but not in the basis of H'_{zz} or H''_{zz}.

Example 8.3   The Hamiltonian

$$H_{xx} = k_x (\sigma_1 \otimes \sigma_1) = \begin{pmatrix} 0 & 0 & 0 & k_x \\ 0 & 0 & k_x & 0 \\ 0 & k_x & 0 & 0 \\ k_x & 0 & 0 & 0 \end{pmatrix}$$

is diagonalized as

$$H_{xx}^D = V_3^{-1} H_{xx} V_3 = \mathrm{diag}(-k_x, -k_x, k_x, k_x) \ ,$$

---

[2]The easiest argument against such a claim is that the sharp operation used in this section is itself unphysical; therefore, all conclusions drawn about channels created with such a sharp operation are unreasonable. Such concerns will be addressed after the discussion of composite Hamiltonians.

with

$$
V_3 = \begin{pmatrix} -1 & 0 & 1 & 0 \\ 0 & -1 & 0 & 1 \\ 0 & 1 & 0 & 1 \\ 1 & 0 & 1 & 0 \end{pmatrix} ,
$$

which leads to

$$
f_\theta = 0
$$

and completely positive reduced dynamics for any value of k_x. A different spectral representation of H_{xx} is found as

$$
H_{xx}^{D\prime} = V_4^{-1} H_{xx} V_4 = \mathrm{diag}(k_x, -k_x, -k_x, k_x) ,
$$

with

$$
V_4 = \begin{pmatrix} 1 & 0 & -1 & 0 \\ 0 & -1 & 0 & 1 \\ 0 & 1 & 0 & 1 \\ 1 & 0 & 1 & 0 \end{pmatrix} .
$$

This second spectral representation of H_{xx} leads to

$$
f_\theta = -\frac{4 k_x t}{\hbar} ,
$$

which has a vanishing negativity if

$$
k_x = -\frac{\pi n \hbar}{2t} \tag{8.5}
$$

in exactly the same manner as H_{zz} in the z-basis.

These observations lead naturally to the question of the basis dependency of the channel negativity.

8.2 $\left(\vec{\tau}, \vec{\tau} \otimes \left(H_d : \vec{\tau} : H_d^\dagger\right), D\right)$ **BASIS DEPENDENCE**

Suppose a state is represented by ρ_φ and the unitary dynamics are represented by U_φ in some basis $\{|\varphi_i\rangle\}$, and we wish to represent them in some new basis $\{|\psi_i\rangle\}$ as ρ_ψ and U_ψ. It can be shown [105] that the two bases are related by some unitary transformation V such that

$$
|\psi_i\rangle = V |\varphi_i\rangle , \quad \langle\psi_i| = \langle\varphi_i| V^\dagger \Rightarrow \rho_\psi = V \rho_\varphi V^\dagger , \quad U_\psi = V U_\varphi V^\dagger
$$

and

$$
|\varphi_i\rangle = V^\dagger |\psi_i\rangle , \quad \langle\varphi_i| = \langle\psi_i| V \Rightarrow \rho_\varphi = V^\dagger \rho_\psi V , \quad U_\varphi = V^\dagger U_\psi V .
$$

If the reduced dynamics are given by

$$
\varepsilon(\rho) = \left(U \rho^\# U^\dagger\right)^\flat
$$

where $U \in \mathcal{B}(\mathcal{H}^{SB})$ is the composite dynamics and $\rho \in \mathcal{S}(\mathcal{H}^S)$ is the initial reduced state, then the reduced dynamics in a new "V basis" would be defined as

$$\left(\left(VUV^\dagger\right)\left(V\rho^\sharp V^\dagger\right)\left(VUV^\dagger\right)^\dagger\right)^\flat = \left(VUV^\dagger V\rho^\sharp V^\dagger VU^\dagger V^\dagger\right)^\flat = \left(VU\rho^\sharp U^\dagger V^\dagger\right)^\flat = \varepsilon_v(\rho)$$

because $V^\dagger V = I$. The entire composite channel is represented in the new V basis except for the partial trace, hence the physical aspects of the channel might seem as if they should remain unchanged.

Notice that

$$
\begin{aligned}
\varepsilon(\vec{\tau}) &= \left(D : \vec{\tau}^\sharp : D^\dagger\right)^\flat \\
&= \left(\begin{pmatrix} 1 & 0 \\ 0 & 0 \end{pmatrix}, \frac{1}{2}\begin{pmatrix} 1 & D_1 D_3^* \\ D_1^* D_3 & 1 \end{pmatrix}, \right. \\
&\qquad \left. \frac{1}{4}\begin{pmatrix} 2 & -i\left(D_1 D_3^* + D_2 D_4^*\right) \\ i\left(D_3 D_1^* + D_4 D_2^*\right) & 2 \end{pmatrix}, \begin{pmatrix} 0 & 0 \\ 0 & 1 \end{pmatrix}\right),
\end{aligned}
$$

where $D = \mathrm{diag}(D_1, D_2, D_3, D_4)$, the sharp operation is defined in Sec. 8.1, and $\vec{\tau}$ is the canonical tomography vector. Similarly,

$$
\begin{aligned}
\varepsilon_v(\vec{\tau}) &= \left(V : D : \vec{\tau}^\sharp : D^\dagger : V^\dagger\right)^\flat \\
&= \left(\begin{pmatrix} 0 & 0 \\ 0 & 1 \end{pmatrix}, \frac{1}{2}\begin{pmatrix} 1 & D_3 D_1^* \\ D_3^* D_1 & 1 \end{pmatrix}, \right. \\
&\qquad \left. \frac{1}{4}\begin{pmatrix} 2 & i\left(D_3 D_1^* + D_4 D_2^*\right) \\ -i\left(D_1 D_3^* + D_2 D_4^*\right) & 2 \end{pmatrix}, \begin{pmatrix} 1 & 0 \\ 0 & 0 \end{pmatrix}\right),
\end{aligned}
$$

where $V = \sigma_1 \otimes \sigma_1$. The first channel has a Choi representation of

$$
C = \begin{pmatrix}
1 & 0 & 0 & x \\
0 & 0 & y & 0 \\
0 & y^* & 0 & 0 \\
x^* & 0 & 0 & 1
\end{pmatrix}
\tag{8.6}
$$

with

$$x = \frac{1}{4}\left(3 D_3 D_1^* + D_4 D_2^*\right)$$

and

$$y = \frac{1}{4}\left(D_1 D_3^* - D_2 D_4^*\right) ,$$

as was already shown in Sec. 8.1. The interesting point to notice is that the Choi representation of the channel in the V basis is

$$
C_v = \begin{pmatrix} 0 & 0 & 0 & y \\ 0 & 1 & x & 0 \\ 0 & x^* & 1 & 0 \\ y^* & 0 & 0 & 0 \end{pmatrix} ,
$$

which has the same spectrum as C. Therefore, if $V = \sigma_1 \otimes \sigma_1$, then the negativity is the same in both the original and the V basis.

Suppose instead that $V = V_1$ from Eq. 8.4. In this case, the Choi representation of the above channel in the V_1 basis is

$$
C_{v1} = \begin{pmatrix} \frac{1}{2} & 0 & -\frac{1}{2} & a \\ 0 & \frac{1}{2} & b & \frac{1}{2} \\ -\frac{1}{2} & b^* & \frac{1}{2} & 0 \\ a^* & \frac{1}{2} & 0 & \frac{1}{2} \end{pmatrix} \tag{8.7}
$$

with

$$
a = \frac{i}{4} \left(D_3 D_2^* - D_1 D_4^* \right)
$$

and

$$
b = \frac{i}{4} \left(D_2 D_3^* - D_4 D_1^* \right) .
$$

The spectrum of C has already been shown in Sec. 8.1. It has also been shown (in that same section) that writing D in polar form such that $D_j = e^{i\theta_j}$ leads to the following condition for the complete positivity of this channel in the original basis (i.e., Eq.8.6):

$$
f_\theta = 2n\pi ,
$$

where $f_\theta = \theta_1 - \theta_2 - \theta_3 + \theta_4$ and n is some integer. The spectrum of C_{v1} is not quite as simple as the spectrum of C, but it can still be written down as follows:

$$
\mathrm{spec}\,(C_{v1}) = \left(\frac{1}{4} \left(2 - \sqrt{2}\sqrt{3 - 4\left|\sin\frac{f_\theta'}{2}\right|} - \cos f_\theta' \right), \right.
$$
$$
\frac{1}{4} \left(2 + \sqrt{2}\sqrt{3 - 4\left|\sin\frac{f_\theta'}{2}\right|} - \cos f_\theta' \right),
$$
$$
\frac{1}{4} \left(2 - \sqrt{2}\sqrt{3 + 4\left|\sin\frac{f_\theta'}{2}\right|} - \cos f_\theta' \right),
$$
$$
\left. \frac{1}{4} \left(2 + \sqrt{2}\sqrt{3 + 4\left|\sin\frac{f_\theta'}{2}\right|} - \cos f_\theta' \right) \right) ,
$$

where $f_\theta' = \theta_1 + \theta_2 - \theta_3 - \theta_4$. Notice that

$$\max \left| \sin \frac{f_\theta'}{2} \right| = 1$$

and is achieved at $f_\theta' = n\pi$ with odd n. The other trigonometric term $\cos f_\theta' = -1$ at $f_\theta' = n\pi$ with odd n, which is the minimum for this term. So, at $f_\theta' = n\pi$ with n odd, the spectrum becomes

$$\mathrm{spec}\,(C_{v1}) = \left(\frac{1}{2}, \frac{1}{2}, -\frac{1}{2}, \frac{3}{2} \right) \ .$$

Notice that

$$\min \left| \sin \frac{f_\theta'}{2} \right| = 0$$

and is achieved at $f_\theta' = n\pi$ with even n, at which point $\cos f_\theta' = 1$, which is its maximum value. So, at $f_\theta' = n\pi$ with n even, the spectrum becomes

$$\mathrm{spec}\,(C_{v1}) = (0, 1, 0, 1) \ .$$

So, this channel can be both completely positive and negative in the V_1 basis. The first eigenvalue is positive whenever

$$2 \geq \sqrt{6 - 8 \left| \sin \frac{f_\theta'}{2} \right| - 2 \cos f_\theta'} \ ,$$

which can be reduced to

$$2 \geq \sqrt{6 - 8 \left| \sin \frac{f_\theta'}{2} \right| - 2 \cos f_\theta'} \ \Rightarrow \ 2 \geq 3 - 4 \left| \sin \frac{f_\theta'}{2} \right| - \cos f_\theta'$$

$$\Rightarrow \ 0 \geq -2 \left| \sin \frac{f_\theta'}{2} \right| + \sin^2 \frac{f_\theta'}{2}$$

$$\Rightarrow \ \sin^2 \frac{f_\theta'}{2} \leq 2 \left| \sin \frac{f_\theta'}{2} \right|$$

where we have used the substitution

$$\cos f_\theta' = 1 - 2 \sin^2 \frac{f_\theta'}{2}$$

by the half-angle formula for sine. Notice that this condition is always satisfied because $\sin \phi \in [-1, 1]$ for $\phi \in \mathbb{R}$, so the first eigenvalue is never negative. This result implies that the only eigenvalue that can be negative is

$$\frac{1}{4} \left(2 - \sqrt{2} \sqrt{3 + 4 \left| \sin \frac{f_\theta'}{2} \right| - \cos f_\theta'} \right) = \frac{1}{4} \left(2 - \sqrt{6 + 8 \left| \sin \frac{f_\theta'}{2} \right| - 2 \cos f_\theta'} \right)$$

which will be positive whenever

$$2 \geq \sqrt{6 + 8\left|\sin\frac{f'_\theta}{2}\right| - 2\cos f'_\theta} \ .$$

This condition can be reduced to

$$2 \geq \sqrt{6 + 8\left|\sin\frac{f'_\theta}{2}\right| - 2\cos f'_\theta} \ \Rightarrow \ 2 \geq 3 + 4\left|\sin\frac{f'_\theta}{2}\right| - \cos f'_\theta$$

$$\Rightarrow \ 0 \geq 2\left|\sin\frac{f'_\theta}{2}\right| + \sin^2\frac{f'_\theta}{2}$$

$$\Rightarrow \ \sin^2\frac{f'_\theta}{2} \leq -2\left|\sin\frac{f'_\theta}{2}\right|$$

where we have again made use of the half-angle formula for sine. The absolute value on the right-hand side and the square on the left-hand side are always positive, so the multiplication by the negative number of the right-hand side implies that this condition is only satisfied when the two sides are equal at

$$\sin\frac{f'_\theta}{2} = 0 \ ,$$

which, as has already been discussed, is achieved at $f'_\theta = n\pi$ with even n. So, this channel is only completely positive in the V_1 basis when

$$f'_\theta = 2n\pi$$

with $n \in \mathbb{Z}$.

This example channel is almost always negative in both the original basis and the V_1 basis, but the conditions for completely positivity in the two bases are different. For the channel to be completely positive in both bases, the channel would have to satisfy the condition $f_\theta = f'_\theta = 2n\pi$. Notice that $f_\theta = f'_\theta$ is not always satisfied. For example, consider the above conditions in terms of the physical parameters of a Hamiltonian. The relationship between θ_j of D and the eigenvalues λ_j of some Hamiltonian H has already been discuss in Sec. 8.1. There it was shown that

$$f_\theta = -\frac{t}{\hbar}\left(\lambda_1 - \lambda_2 - \lambda_3 + \lambda_4\right) \ ,$$

and it follows that

$$f'_\theta = -\frac{t}{\hbar}\left(\lambda_1 + \lambda_2 - \lambda_3 - \lambda_4\right) \ .$$

Consider the example Hamiltonian $H_{zz} = \mathrm{diag}(k_z, -k_z, -k_z, k_z)$ from Sec. 8.1. This Hamiltonian leads to

$$f_\theta = -\frac{4k_z t}{\hbar}$$

(as was already shown), and

$$f_\theta' = 0 .$$

This result means that the example channel has a basis dependent negativity if the composite dynamics are defined as $D = \exp\{(-it/\hbar)H_{zz}\}$. This channel will be completely positive in the original basis when $k_z = -\pi n \hbar/(2t)$ (as was already discussed in Sec. 8.1), but it will *always* be completely positive in the V_1 basis.

It is possible to find examples of D with the same negativity in both the original and the V_1 basis. For example, consider the controlled-phase gate (which will be discussed in much more detail in later sections) which is defined as

$$D = \begin{pmatrix} 1 & 0 & 0 & 0 \\ 0 & 1 & 0 & 0 \\ 0 & 0 & 1 & 0 \\ 0 & 0 & 0 & -1 \end{pmatrix} .$$

These composite dynamics lead to $\theta_1 = \theta_2 = \theta_3 = 2n\pi$ and $\theta_4 = (2n+1)\pi$ with $n \in \mathbb{Z}$, which implies $f_\theta = \pi$ and $f_\theta' = -\pi$. So, in the original basis, the spectrum of the Choi representation is

$$\mathrm{spec}\,(C) = \left(\frac{1}{2}, \frac{3}{2}, -\frac{1}{2}, \frac{1}{2}\right)$$

and in the V_1 basis, the spectrum of the Choi representation is

$$\mathrm{spec}\,(C) = \left(\frac{1}{2}, \frac{1}{2}, -\frac{1}{2}, \frac{3}{2}\right) .$$

It follows that the negativity $\eta = 1/6$ in both bases. Hence, there are some composite dynamics (and some bases) for which this example channel has a basis independent negativity. But, an example of basis dependent negativity was also given in this section. So, the final conclusion must be that negativity is, in general, dependent on the basis when the partial trace is left unchanged during a basis rotation. Such a result is perhaps unsurprising given that rotating the partial trace along with the rest of the channel parameters is required in a "true" basis change of the channel.

Consider a channel defined by some tomography vector $\vec{\tau}$, some sharp operation $\sharp$, and some composite dynamics U such that the channel is not always completely positive. This channel can be written down as

$$\varepsilon(\vec{\tau}) = \left(U : \vec{\tau}^\sharp : U^\dagger\right)^\flat .$$

Suppose now that the composite dynamics and initial composite state are represented in a new basis given by $U^\dagger$. The matrix U represents the composite dynamics, but it can also be thought of as a change of basis because it is a unitary transformation. The channel in this new basis will be

$$
\begin{aligned}
\varepsilon_U(\vec{\tau}) &= \left(\left(U^\dagger U U\right):\left(U^\dagger:\vec{\tau}^\sharp:U\right):\left(U^\dagger U^\dagger U\right)\right)^\flat \\
&= \left(\left(U^\dagger U U U^\dagger\right):\vec{\tau}^\sharp:\left(U U^\dagger U^\dagger U\right)\right)^\flat \\
&= \left(\vec{\tau}^\sharp\right)^\flat \\
&= \vec{\tau}
\end{aligned}
$$

by the unitarity of U and the consistency of the sharp operation. The unitarity of U implies $U^\dagger U U U^\dagger = U U^\dagger U^\dagger U = I$ where I is the identity operator, and the identity operator can always be written in the local unitary form of $I = \tilde{I} \otimes \tilde{I}$ where $\tilde{I}$ are identity operators that act on smaller spaces than the original I. Local unitary composite dynamics always lead to completely positive reduced dynamics. So, there is always a basis for the composite dynamics and state where a negative channel ε is completely positive and it is the basis defined by $U^\dagger$ where U is the composite dynamics.

As a quick aside, notice that the original channel ε in the above example was defined to be negative, so it might have had some positivity domain that was not all of $S(\mathcal{H}^S)$. But, the rotated channel ε_U is completely positive, so has a positivity domain that is all of $S(\mathcal{H}^S)$.[3] It is important to notice here that the positivity domain is also basis dependent when the partial trace is not rotated.

The statement of "basis dependence" of the kind discussed in the section thus far (i.e., with the partial trace in the original, un-rotated basis) is very much a statement of the channel dependence of the negativity. In the first example, the channel ε is defined by the partial trace of the composite channel, i.e.,

$$
\varepsilon(\rho) = (C(\rho))^\flat
$$

where the composite channel is $C(\rho) = U\rho^\sharp U^\dagger$. The channel in the new V basis is just the partial trace of the conjugated composite channel, i.e.,

$$
\varepsilon_v(\rho) = \left(V C(\rho) V^\dagger\right)^\flat \ ,
$$

but the partial trace function is exactly the same. The channel in the new basis could be written as

$$
\varepsilon_v(\rho) = (C'(\rho))^\flat \ ,
$$

with $C'(\rho) = \tilde{U}\rho^\sharp\tilde{U}^\dagger$ and $\tilde{U} = VU$, which makes it clear that the channel in the new basis could be considered a completely new channel. As such, this statement of the basis dependence of the negativity is a statement of the channel dependence of the negativity.

---

[3]This can be seen immediately in the above example because $\vec{\tau}$ is a tomography vector and, as such, is a basis for $S(\mathcal{H}^S)$. Process tomography on the channel $\varepsilon_U(\rho)$ for some given state ρ would simply be state tomography on ρ.

It is physically unreasonable to use the same partial trace function in a discussion of the basis dependence. For example, if the composite dynamics are the identity operator, then tracing out the bath will simply yield the initial reduced system state. If a basis change is defined by the SWAP operation (i.e., the operation that swaps the states of the bath and the reduced system), then "tracing out the bath" in the new basis will yield the initial state of the bath. If, however, the partial trace operation itself is written in the new basis such that "tracing over the bath" becomes "tracing over the reduced system" in the new basis, then the behavior of this channel is basis independent. Rotating to a new basis should not provide the experimenter with previously inaccessible information.

Defining the partial trace in such a way that the reduced dynamics (and hence, the negativity) are basis independent is equivalent to rotating the observer to the new basis. The "basis change" discussed in this section is more correctly thought of as a "channel change." The new channel is related to the old channel because the composite channels are related by a basis change. The flat operator (i.e., the partial trace operator) defines the experimenter and the system-bath divide. As such, it must be rotated, along with everything else in the channel definition, if the channel is represented entirely in a new basis.

At this point in the discussion, it suffices to recognize that negativity is a property of a very specific channel, and it is basis independent (with the understanding that care must be taken to define the partial trace in the new rotated basis).

CHAPTER 9

Rabi Channels

The negativity has thus far been a mathematical concept with, at best, a nebulous physical interpretation as some measure of the system-bath correlation and coupling. It is important to consider examples of negative channels in situations where there is already some established physical understanding of the system so we can attempt to gain some physical intuition about the negativity. To that end, we will be discussing negative channels using the Rabi model of a two level atom. This discussion necessarily begins with a brief introduction to the Rabi model.

9.1 RABI MODEL

We will refer to the following Hamiltonian as the "Rabi model:"

$$H_q = \frac{\hbar}{2} \begin{pmatrix} -\nu & \Omega \\ \Omega & \nu \end{pmatrix} ,$$

where $\nu \in \mathbb{R}$ is called the detuning and $\Omega \in \mathbb{R}$ is called the Rabi frequency. A derivation of the Hamiltonian can be found in Appendix A.5.

This model is commonly used in theoretical descriptions of nuclear magnetic resonance (NMR) and other quantum information experiments [67], but this model was original developed in quantum optics as a way to understand the behavior of an ideal two-level atom interacting with a classical electromagnetic field. Derivations of this model using "semiclassical methods" can be found in [6, 51, 60, 62] and using perturbation theory in [60, 70, 95]. It is possible to verify this model using a fully quantum treatment of the interacting electromagnetic field [6], but such discussions are well outside the scope of the present discussion. The important issue for us is that the Rabi model accurately describes the behavior of several different physical implementations of qubit systems [67].

One reason the Rabi model is a popular model for the qubit is because the eigenvalues and eigenvectors of the Hamiltonian can be written down as simple functions of the Hamiltonian parameters. The eigenvalues and eigenvectors are [51]

$$\lambda_\pm = \pm \frac{\hbar}{2} \sqrt{\nu^2 + \Omega^2}$$

with

$$|\lambda_+\rangle = \cos\left(\frac{\theta}{2}\right) |1\rangle + \sin\left(\frac{\theta}{2}\right) |0\rangle ,$$

$$|\lambda_-\rangle = -\sin\left(\frac{\theta}{2}\right)|1\rangle + \cos\left(\frac{\theta}{2}\right)|0\rangle \ ,$$

where θ is called the mixing angle and is defined as

$$\theta = \arctan\left(\frac{\Omega}{\nu}\right) \ .$$

Notice that these states are superpositions of the energy states of the atom (i.e., $|0\rangle$ and $|1\rangle$). These new states are referred to as "dressed states" because the photon from the electromagnetic field has "dressed" the energy levels of the atom [95].

Investigating the dynamics of the atom and external field is straightforward in the basis of the dressed states. Dressed states are often used to understand the interaction of a laser with the "bare states" of the atom [6]. Unfortunately, it is often difficult to find the dressed states when the atom is subject to interactions other than the classical monochromatic field we have used in the derivation of this Rabi model [6]. In particular, the situation becomes much more difficult when there is coupling present between the bare states of the atom. This issue will be addressed in Sec. 9.3. The difficulty of finding dressed states in general is a big motivation for why we will be working exclusively in the bare state basis. The bare states also have straightforward interpretations (i.e., they are the unperturbed energy levels of the two level atom) which will help with the interpretation of the negativity results that follow. This basis choice for the composite dynamics implies that the tomography vector, the sharp operation, and the partial trace are all likewise defined using the bare basis.

We will be using the Rabi Hamiltonian to model the reduced system qubit, and the bath qubit, in an effort to understand negativity in terms of the physical parameters of the Hamiltonian. A *Rabi channel* will refer to the single qubit channel where the bath qubits and the reduced system are all described by individual Rabi models. The simplest such channel will have a single qubit bath and no coupling terms in the composite Hamiltonian, and this channel will be discussed in the next subsection.

9.2 RABI CHANNELS WITH NO COUPLING

Suppose the universe consists of two Rabi atoms. One atom will be the reduced system and the other will be the bath. The composite dynamics of this universe can be understood in terms of the composite Hamiltonian

$$H_u = \frac{\hbar}{2}\begin{pmatrix} -\nu_s & \Omega_s \\ \Omega_s & \nu_s \end{pmatrix} \otimes I + I \otimes \frac{\hbar}{2}\begin{pmatrix} -\nu_b & \Omega_b \\ \Omega_b & \nu_b \end{pmatrix} \equiv H_s \oplus H_b \ ,$$

where the kronecker sum $\oplus$ is defined by the action $A \oplus B = A \otimes I + I \otimes B$ with I being the identity operator on the appropriate space. The subscripts s and b refer to the parameters restricted to the reduced system and bath, respectively.

The kronecker sum obeys the following nice property for matrix exponentials [44]:

$$e^{A \oplus B} = e^A \otimes e^B \ .$$

This property significantly reduces the difficulty of finding the unitary evolution of our example universe. The composite unitary evolution is given as

$$
\begin{aligned}
U^{SB} &= \exp\left(\frac{-it}{\hbar} H_u\right) \\
&= \exp\left(\frac{-it}{\hbar}(H_s \oplus H_b)\right) \\
&= \exp\left(\frac{-it}{\hbar} H_s\right) \otimes \exp\left(\frac{-it}{\hbar} H_b\right) \\
&\equiv U^S \otimes U^B
\end{aligned}
$$

with

$$U^S = \exp\left(\frac{-it}{\hbar} H_s\right) \in \mathcal{B}\left(\mathcal{H}^S\right)$$

and

$$U^B = \exp\left(\frac{-it}{\hbar} H_b\right) \in \mathcal{B}\left(\mathcal{H}^B\right) \ .$$

Therefore, the composite evolution (in the original basis of H_u) can written down in a local unitary form. As was already shown, this result implies that the reduced system dynamics always have a vanishing negativity.

9.3 RABI CHANNELS WITH COUPLING

Suppose the two atoms in the Rabi channel example of the last subsection were coupled in some way. The physical mechanism for their coupling will not be important to this discussion, but the coupling terms in the Hamiltonian will change the behavior of the negativity. The negativity will now depend on the coupling as well as the energy of the composite system and the initial rotation of the bath qubit.

The channel will be governed by the Hamiltonian

$$H_u = \frac{\hbar}{2}\begin{pmatrix} -v_s & \Omega_s \\ \Omega_s & v_s \end{pmatrix} \otimes I + I \otimes \frac{\hbar}{2}\begin{pmatrix} -v_b & \Omega_b \\ \Omega_b & v_b \end{pmatrix} + H_C \ ,$$

where I is the identity operator and H_C describes the coupling between the two atoms. The coupling Hamiltonian acts on both atoms and can be written in general as

$$H_C = \sum_{ij} k_{ij}\sigma_i \otimes \sigma_j \ ,$$

where k_{ij} are constants and the σ_i operators have already been introduced. For simplicity, consider

$$H_C = k_z \sigma_3 \otimes \sigma_3 \ ,$$

where $k_z = k_{33}$ is some positive real number. Assume that the atoms are identical and subject to the same classical field. These assumptions lead to

$$H'_u = H_q \otimes I + I \otimes H_q + k_z \sigma_3 \otimes \sigma_3 \ , \tag{9.1}$$

where H_q is

$$H_q = \frac{\hbar}{2} \begin{pmatrix} -\nu & \Omega \\ \Omega & \nu \end{pmatrix} \ .$$

The composite dynamics are then

$$U = \exp \left(\frac{-it}{\hbar} H'_u \right) \ .$$

The next step to defining this channel is the define the sharp operation and the tomography vector. The sharp operation will be defined as

$$\tau_i^{\#} = \tau_i \otimes \left(R(\phi) \tau_i R(\phi)^\dagger \right) \ , \tag{9.2}$$

where

$$R(\phi) = \begin{pmatrix} \cos \phi & -\sin \phi \\ \sin \phi & \cos \phi \end{pmatrix} \tag{9.3}$$

with $\phi \in \mathbb{R}$ and $\vec{\tau}$ is the canonical tomography vector. This sharp operation is sufficiently general to study the effects of the sharp operation on the negativity, and it stills meets the requirements discussed in Sec. 1.4.3.

This channel is not as simple as the previous examples. The channel is dependent on five separate parameters: the detuning of the two atoms ν, the Rabi frequency of the two atoms Ω, the coupling between the two atoms k_z, the initial rotation of the bath atom ϕ, and time t. Such a 5-dimensional parameter space can make the Choi representation of the channel difficult to interpret directly.

This example can be simplified with a few assumptions about the system. The field is assumed to be resonant with the atoms, i.e., $\nu = 0$. All of the parameters are scaled such that $\Omega = 1$ with everything in units of $\hbar = 1$. The initial rotation of the bath atom will be fixed by a Hadamard rotation rather than more general $R(\phi)$. These assumptions will reduce the 5-dimensional parameter space to a much simpler 2-dimensional parameter space of (k_z, t).

Figure 9.1 is the behavior of the negativity as a function of time if the coupling coefficient is fixed at $k_z = \pi/2$. Figure 9.2 is the behavior of the negativity as a function of the coupling coefficient if the time is fixed at $t = \pi/2$. Both plots show the channel negativity η calculated

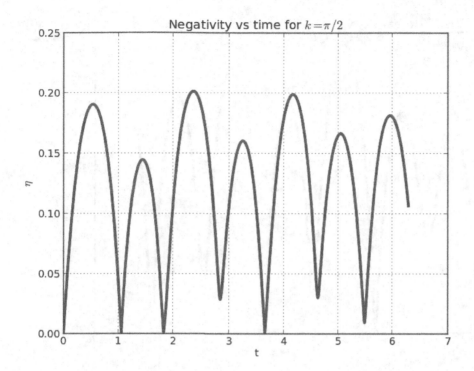

Figure 9.1: The negativity η only depends on the elapsed time t when the coupling constant k_z is fixed. This plot is for a fixed coupling coefficient of $k_z = \pi/2$ (and given the assumptions discussed in the text).

with one parameter (either t or k_z) over the range $[0, 2\pi]$ while holding the other parameter fixed. Notice that the negativity is only zero (i.e., the channel is only completely positive) at a small number of fixed points.

The negativity can be plotted as a function of both the coupling coefficient and time for a visualization of the 2-dimensional parameter space over the range $[0, \pi]$. Figure 9.3 is that plot.

This more complicated example shows that the negativity is dependent on the physical parameters of the Hamiltonian. As such, a measurement of the negativity (through a tomography experiment) will yield information about parameters that might be inaccessible directly, e.g., the coupling k_z between the reduced system and bath qubits in this Rabi channel.

The double Rabi atom universe presented here shows that theoretically negative channels can be found even with experimentally motivated Hamiltonians.

Is it possible to create negative channels in the lab? This question is of paramount impor-tance. The theoretical work presented so far seems to imply such a claim, but an experiment that

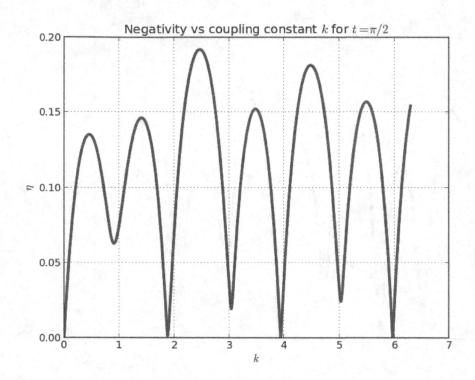

Figure 9.2: The negativity η only depends on the coupling constant k_z when the elapsed time t is fixed. This plot is for a fixed time of $t = \pi/2$. (See the text for a discussion of the other assumptions.)

measures the negativity of some channel in the lab (and compares it to the theoretical predictions presented here) is the only way to really answer this question. Before exactly such an experiment is proposed, an important theoretical issue needs to be addressed: Every channel presented so far has depended on the sharp operation $\vec{\tau}^\sharp = \vec{\tau} \otimes R : \vec{\tau} : R^\dagger)$ where R is some unitary rotation. Is such a sharp operation physically reasonable? Is it possible such a sharp operation might occur in nature? Both of these questions will be answered in the next section.

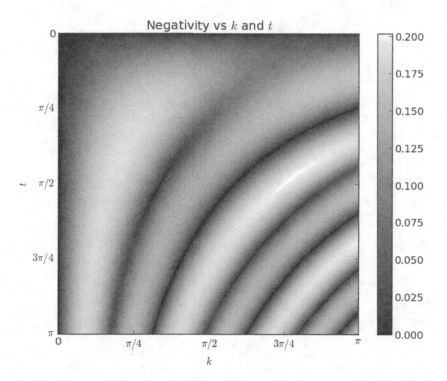

Figure 9.3: The negativity η of the example Rabi channel in the text depends on both the coupling constant k_z and the elapsed time t. Notice that the channel is almost always negative, and the negativity appears cyclic. The text discusses the assumptions used to produce this plot.

CHAPTER 10

Physical Motivations for Sharp Operations

Sharp operations can be thought of as a by-product of the preparation procedure. Every quantum experiment must begin with a preparation in some way at some point in time, and preparing the reduced system will only leave the bath completely unaffected if the reduced system is isolated. In such cases, there is no need to discuss baths, complete positivity, or sharp operations.

Sharp operations of the form described here (i.e., linear and consistent, but not necessarily positive) have been discussed in [77] under the more typical name of assignment maps. In that work, sharp operations of this form are proven to be Hermiticity- and trace-preserving. In [54], it is argued that the reduced dynamics described throughout this work should be thought of more as theoretical tools that are not necessarily compatible with the "process map" output of tomography experiments. A process map is the superoperator found experimentally with a tomography experiment, and the reduced dynamics (which include the sharp operation) form a "dynamical map" that will have some positivity domain that may depend on the initial correlations of the reduced system and bath. This is the key idea here. In the language of [54], a process map and a dynamical map both describe the same physical process only if the initial correlations are correctly accounted for in the sharp operation. For example, the theoretical description of such channels ("dynamical maps") will coincide with experimental reality ("process maps") only if the sharp operation accurately describes the initial correlation between the reduced system and the bath. This concern is precisely why it is important to determine if the sharp operations we use can be created in the lab.

It might be argued that the example sharp operation of

$$\tau_i = \tau_i \otimes \left(H_d \tau_i H_d^\dagger \right) \ , \tag{10.1}$$

where τ_i is the ith states of the tomography vector $\vec{\tau}$ and H_d is the Hadamard operator (Sec. 8.1 shows this sharp operation in use), does not have a simple physical interpretation. To address such concerns, consider the situation arising when the reduced system and the bath are initially entangled in the state

$$|\Psi\rangle = \frac{|0+\rangle + |1-\rangle}{\sqrt{2}} \ .$$

Preparing the tomography states on the reduced systems yields

$$
\begin{aligned}
|0\rangle\langle 0| \otimes I |\Psi\rangle &= |0+\rangle \\
|1\rangle\langle 1| \otimes I |\Psi\rangle &= |1-\rangle \\
|+\rangle\langle +| \otimes I |\Psi\rangle &= |+0\rangle \\
|+_i\rangle\langle +_i| \otimes I |\Psi\rangle &= |+_i\rangle (|+\rangle - i |-\rangle),
\end{aligned}
$$

up to a normalization factor.[1] These states do not exactly correspond to the sharp operation of Eq. 10.1. Define a new sharp operation on $\vec{\tau}$ using the above projective measurements, i.e.,

$$
\tau_i^{\#_p} = \frac{(\tau_i \otimes I) |\Psi\rangle\langle\Psi| (\tau_i \otimes I)}{\mathrm{Tr}\left((\tau_i \otimes I) |\Psi\rangle\langle\Psi|\right)} \ . \tag{10.2}
$$

Notice $\tau_i^{\#} = \tau_i^{\#_p}$ for $i \in \{1, 2, 4\}$ but not for $i = 3$. However, $\left(Z_c \tau_i^{\#} Z_c^{\dagger}\right)^{\flat} = \left(Z_c \tau_i^{\#_p} Z_c^{\dagger}\right)^{\flat} \ \forall i$. This observation makes it clear that the Choi matrix for a single qubit channel with composite dynamics described by Z_c (i.e., Eq. 12.1) will be the same regardless of whether the sharp operation is described bb Eq. 10.1 or Eq. 10.2.

The above sharp operation can be explained as the reduced system and the bath initially being entangled in the pure state $|\Psi\rangle$ and then ideally preparing a tomography state on the reduced system. The appearance of negative channels after projective measurements of reduced systems entangled with the bath is currently an active area of research [98].

A sharp operation does not always need to be explained with prior entanglement between the reduced system and the bath. As pointed out in [54], the two main preparation procedures in quantum mechanics are the "preparation by measurement" described above and "stochastic preparations" [54]. Stochastic preparations are preparations of a system by setting some macroscopic parameter (e.g., the temperature or external magnetic field) and letting the system reach some equilibrium state. For example, if the qubit is represented by the spin of an electron in a quantum dot, then preparation of only that spin by cooling would require the spatial extent of the cooling to be bounded by the spatial extent of the quantum dot. This experimental requirement might not be met; the entire sample containing the quantum dot might be cooled resulting in every other spin in neighboring quantum dots to be similarly be prepared. Every qubit would be prepared identically because every qubit is represented by a spin in the same (presumably spatially uniform) temperature field. This situation is exactly described by the sharp operation

$$
\rho^{\#} = \rho \otimes \rho \otimes \rho \otimes \cdots \ .
$$

This situation is a simple example of a sharp operation which requires no prior entanglement between the reduced system and bath.

The conclusion is that sharp operations can be implemented in the lab as preparation procedures; as such, sharp operations are physical. If quantum operations must be completely positive

---

[1]The normalization term is left out of the expressions in this section for convenience and because it is not relevant to the argument.

to be considered physically reasonable and if the controlled-phase gate and sharp operation of Eq. 10.1 are both physically reasonable, then what part of the evolution described by Eq. 12.1 is not physically reasonable? The partial trace seems to be the only other major element of the mathematical description of that channel and the partial trace operation is well established in physics as a linear and positive operation [23] with a clear physical interpretation involving the consistency of expectation values in extended Hilbert spaces [28].

Sharp operations represent preparation procedures in the open systems setting. Negative channels can arise if the preparation procedure is not perfect or if the reduced system and bath are entangled prior to the preparation procedure. Both situations can (and probably do) happen in nature. It will be shown in later sections that a controlled "bath" will allow an experimenter to engineer a desired sharp operation and experimentally measure the negativity of a channel.

CHAPTER 11

Negative Qubit Channel Examples with Multi-Qubit Baths

The example channel $\left(\vec{\tau}, \vec{\tau} \otimes \left(H_d : \vec{\tau} : H_d^\dagger\right), D\right)$ can be extended by redefining the 4×4 matrix D as an $2^M \times 2^M$ matrix D'. This new channel describes a qubit channel with a bath of $M - 1$ qubits as opposed to the single qubit bath of all the previous examples. The larger bath will also require a different sharp operation. Suppose

$$\vec{\tau}^\# = \vec{\tau} \otimes \left(H_d : \vec{\tau} : H_d^\dagger\right) \otimes \bigotimes_{i=3}^{M} |0\rangle\langle 0| \ ,$$

i.e., the first bath qubit will act exactly as the only bath qubit of the previous examples and the rest of the bath qubits will simply be prepared in a fixed state of $|0\rangle\langle 0|$. This sharp operation could arise, for example, from a combination of the stochastic and measurement preparation methods described in the previous section. The entire system might be cooled to the ground state $|0\rangle\langle 0|$ except for two qubits that somehow manage to remain maximally entangled. Preparation of the reduced system in the "preparation by measurement" manner described in the previous section could then result in this sharp operation. Notice

$$\varepsilon(\rho) \equiv \left(D' \rho^\# D'^\dagger\right)^\flat = \begin{pmatrix} x & y \\ y^* & 1 - x \end{pmatrix} \ ,$$

where

$$x = \sum_{i=1}^{2^{M-1}} \rho_{ii}^\#$$

and

$$y = \sum_{i=1}^{2^{M-1}} \sum_{j=(2^{M-1})+1}^{2^M} D_i' \rho_{ij}^\# D_j'^* \ .$$

The complex number D_i' is the ith diagonal element of D' and ρ is some valid density matrix. This is the straightforward extension of Eq. 8.1 to a multi-qubit bath. The transformation matrix

$\hat{R}$ from that section can be used to find

$$\varepsilon(\vec{\tau}) = \left(\begin{pmatrix} 1 & 0 \\ 0 & 0 \end{pmatrix}, \right.$$
$$\frac{1}{4} \begin{pmatrix} 0 & D'_s D'^*_1 - D'_{s+r} D'^*_{1+r} \\ 3D'_1 D'^*_s + D'_{1+r} D'^*_{s+r} & 0 \end{pmatrix},$$
$$\frac{1}{4} \begin{pmatrix} 0 & 3D'^*_1 D'_s + D'^*_{1+r} D'_{s+r} \\ D'^*_s D'_1 - D'^*_{s+r} D'_{1+r} & 0 \end{pmatrix},$$
$$\left. \begin{pmatrix} 0 & 0 \\ 0 & 1 \end{pmatrix} \right),$$

with $s = (2^{M-1}) + 1$ and $r = 2^{M-2}$, which leads to a Choi representation of this channel of

$$C_{D'} = \mathbf{C} \odot \left(D' : \vec{\tau}^{\#} : D'^{\dagger} \right)^{\flat} = \begin{pmatrix} 1 & 0 & 0 & m \\ 0 & 0 & n & 0 \\ 0 & n^* & 0 & 0 \\ m^* & 0 & 0 & 1 \end{pmatrix} \tag{11.1}$$

with

$$m = 3D'^*_1 D'_s + D'^*_{1+r} D'_{s+r}$$

and

$$n = D'^*_s D'_1 - D'^*_{s+r} D'_{1+r} \ .$$

The spectrum of this channel would be

$$\operatorname{spec}(C_{D'}) = \left(1 - \sqrt{mm^*}, 1 + \sqrt{mm^*}, -\sqrt{nn^*}, \sqrt{nn} \right) ,$$

with

$$\pm\sqrt{mm^*} = \pm\frac{1}{\sqrt{8}}\sqrt{5 + 3\cos(f_\nu t)}$$

and

$$\pm\sqrt{nn^*} = \pm\frac{1}{2}\sin\left(\frac{f_\nu t}{2}\right) \ .$$

The argument f_ν is defined in terms of the composite Hamiltonian that generates D', similar to Eq. 8.2, i.e.,

$$f_\nu = \nu_1 - \nu_{1+r} - \nu_s + \nu_{s+r} \ ,$$

where ν_i is the ith eigenvalue of the composite Hamiltonian. Notice (with $\hbar = 1$)

$$f_\nu t = 2\pi n \Rightarrow \eta_{D'} = 0$$

where $n \in \mathbb{Z}$, t is the elapsed time defining D' and $\eta_{D'}$ is the negativity of this example M qubit channel. This channel is, like most of the previous examples, almost always negative. The addition of a multi-qubit bath does not force complete positivity.

In general,

$$\rho^\sharp = \rho \otimes b = \begin{pmatrix} \rho_{11}b & \rho_{12}b \\ \rho_{21}b & \rho_{22}b \end{pmatrix} \tag{11.2}$$

where $\rho \in \mathcal{S}(\mathcal{H}^S)$ is the reduced system state (a single qubit) and $b \in \mathcal{S}(\mathcal{H}^B)$ is some bath state (not necessarily a single qubit), and

$$\begin{aligned}
D\rho^\sharp D^\dagger &= D\left(\rho \otimes b\right) D^\dagger \\
&= \begin{pmatrix} D_1 & 0 & \cdots & 0 \\ 0 & \ddots & \ddots & \vdots \\ \vdots & \ddots & \ddots & 0 \\ 0 & \cdots & 0 & D_{2M} \end{pmatrix} \begin{pmatrix} \rho_{11}b & \rho_{12}b \\ \rho_{21}b & \rho_{22}b \end{pmatrix} \begin{pmatrix} D_1^* & 0 & \cdots & 0 \\ 0 & \ddots & \ddots & \vdots \\ \vdots & \ddots & \ddots & 0 \\ 0 & \cdots & 0 & D_{2M}^* \end{pmatrix}.
\end{aligned}$$

These equations imply

$$x = \sum_{i=1}^{2^{M-1}} \rho_{ii}^\sharp = \rho_{11} \operatorname{Tr}(b) = \rho_{11}$$

and

$$y = \sum_{i=1}^{2^{M-1}} \sum_{j=(2^{M-1})+1}^{2^M} D_i' \rho_{ij}^\sharp D_j'^* = \sum_{i=1}^{2^{M-1}} \sum_{j=(2^{M-1})+1}^{2^M} \rho_{12} D_i' b_{ii} D_j'^* = \kappa_b \rho_{12} \ ,$$

with $\kappa_b = \sum_{i=1}^{2^{M-1}} \sum_{j=(2^{M-1})+1}^{2^M} D_i' b_{ii} D_j'^*$. This implies the reduced dynamics on the canonical tomography basis can be written down as

$$\varepsilon(\vec{\tau}) = \left(\begin{pmatrix} 1 & 0 \\ 0 & 0 \end{pmatrix}, \frac{1}{2}\begin{pmatrix} 1 & \kappa_+ \\ \kappa_+^* & 1 \end{pmatrix}, \frac{1}{2}\begin{pmatrix} 1 & -i\kappa_{+i} \\ i\kappa_{+i}^* & 1 \end{pmatrix}, \begin{pmatrix} 0 & 0 \\ 0 & 1 \end{pmatrix} \right) \ ,$$

where κ_+ is a function of the bath state b_+ resulting from the sharp operation acting on $|+\rangle\langle+|$ and κ_{+i} is a function of some other bath state b_{+i} resulting from the sharp operation acting on $|+_i\rangle\langle+_i|$. The Choi representation of this channel is of the form of Eq. 11.1 with

$$m = \frac{\kappa_+ + \kappa_{+i}}{2}$$

and

$$n = \frac{\kappa_+^* - \kappa_{+i}^*}{2} \ .$$

This channel will be negative when $nn^* \neq 0$.

The evolution of a single qubit in a sea of qubits will, presumably, be a common problem in the world of quantum technologies. The reduced system qubit is not required to be initially correlated with all of the bath qubits or even some fixed majority of them. The qubit channel will almost always be negative if the reduced system qubit is initially correlated to a single member of the M qubit bath.

This problem becomes much more difficult when considering continuous baths, as is usually the case in the study of open quantum systems. In general, the Liouville-von Neumann evolution of the composite system will always lead to reduced system dynamics independent of complete positivity assumptions, i.e.,

$$\dot{\rho} = -\frac{i}{\hbar} \left([H^{SB}, \rho^{\sharp}] \right)^{\flat}$$

describes the reduced system dynamics where $H^{SB} \in \mathcal{B}(\mathcal{H}^{SB})$ is the composite Hamiltonian governing the composite dynamics and $\rho \in \mathcal{S}(\mathcal{H}^S)$ is the state of the reduced system. This equation does not require any assumptions of complete positivity.

Unfortunately, this equation is also unwieldy and resistant to analytical analysis. As a result, several assumptions are typically used to derive simpler evolution equations. The Markovian master equation, also-called the "Lindblad" or "Kossakowski-Lindblad" equation, is an example of such a simplification, and it takes the form

$$\mathcal{L}\rho^S(t) = -i[H, \rho(t)] + \sum_{k=1}^{N^2-1} \gamma_k \left(A_k \rho^S(t) A_k^\dagger - \frac{1}{2} A_k^\dagger A_k \rho^S(t) - \frac{1}{2} \rho^S(t) A_k^\dagger A_k \right) , \quad (11.3)$$

where $\gamma_k \geq 0 \ \forall k$ and the operators A_k are called "Lindblad" operators. The operator H will generate the unitary (i.e., Hamiltonian) dynamics of the evolution, but in general, H is not equal to the system Hamiltonian, and in the literature, H is usually referred to as the "Hermitian part" of the system Hamiltonian. The derivation of this equation can be found in [18]. The important points for this discussion are the assumptions that lead to this equation.

The first key assumption in the derivation of Eq. 11.3 is the semigroup property. Suppose the dynamics of the reduced system from times $t = 0$ to t are represented as

$$\rho^S(t) = V(t)\rho^S(0) .$$

The map

$$V(t) : \mathcal{S}(\mathcal{H}^S) \to \mathcal{S}(\mathcal{H}^S)$$

is called a "dynamical map." The dynamical map $V(t)$ is defined for a fixed time $t \geq 0$, and allowing t to vary produces the one-parameter family $\{V(t)|t \geq 0\}$ of dynamical maps which completely describe the future time evolution of the open system. Here, it is assumed that memory effects of the bath are negligible, i.e., the evolution is Markovian. This idea is made formal with the semigroup property

$$V(t_1)V(t_2) = V(t_1 + t_2), \quad t_1, t_2 \geq 0 .$$

The introduction of this rule creates a semigroup from the one-parameter family $\{V(t)|t \geq 0\}$ of dynamical maps. The semigroup of dynamical maps has a multiplication operation among the group elements defined by the above Markov rule but each group element does not necessarily have an inverse, hence the collection of dynamical maps only forms a *semi*group.

Given a semigroup of dynamical maps, there must be some linear operator $\mathcal{L}$ that will allow the semigroups to be represented in exponential form as

$$V(t) = e^{\mathcal{L}t} \ .$$

This description of the dynamical map $V(t)$ in terms of some generator $\mathcal{L}$ might seem to come out of the blue, but it is, in fact, a well-established property of dynamical semigroups. The reference [32] gives all the gory details, but a brief explanation can be seen as follows. Remember the dynamical semigroup is defined by the semigroup property $V(t_1)V(t_2) = V(t_1 + t_2)$ for $t_1, t_2 \geq 0$. Above, $\rho(t) = V(t)\rho(0)$ was given as a definition of the map $V(t)$. Hence, $\rho(0) = V(0)\rho(0)$ or

$$V(0) = I \ .$$

This initial value and the semigroup property actually form the meat of a famous functional equation posed by Cauchy in 1821. Cauchy wanted to find all maps $T(\cdot) : \mathbb{R}_+ \to \mathbb{C}$ that satisfy the functional equation

$$T(t + s) = T(t)T(s) \ \ \forall \ t, s \geq 0 \ ,$$

with

$$T(0) = 1 \ .$$

The exponential functions solve this functional equation, i.e.,

$$T(t) = e^{t\alpha}$$

for any $\alpha \in \mathbb{C}$. It turns out that all solutions of this functional equation will have this form [32].

Notice the definition $T(t) := e^{t\alpha}$ for some $\alpha \in \mathbb{C}$ and all $t \geq 0$ implies $T(\cdot)$ is differentiable and satisfies the differential equation

$$\frac{d}{dt}T(t) = \alpha T(t) \ ,$$

with the initial value $T(0) = 1$. The converse is also true: the function $T(\cdot) : \mathbb{R}_+ \to \mathbb{C}$ defined by $T(t) = e^{t\alpha}$ for some $\alpha \in \mathbb{C}$ and all $t \geq 0$ are the only functions that satisfy the above differential equation. The proof of this proposition and its converse can both be found in [32].

The semigroup structure of dynamical maps comes about from basic physical assumptions, i.e., the assumption of Markovian evolutions and the ability to compose dynamical maps. Any family of maps obeying the given functional equation can be rewritten in terms of a generator obeying a differential equation. Hence, these quantum dynamical maps can be thought of in terms of a generator $\mathcal{L}$ (called the "Liouvillian"). Dealing with the generator rather than the maps themselves is the great power of quantum dynamical semigroup theory. As Engel says [32]:

"...in most cases a complete knowledge of the maps $T(\cdot)$ is hard, if not impossible, to obtain. It was one of the great discoveries of mathematical physics, based on the invention of calculus, that, as a rule, it is much easier to understand the 'infinitesimal changes' occurring at any given time. In this case, the system can be described by a differential equation replacing the functional equation..."

The derivation of the generator $\mathcal{L}$ of the dynamical maps $V(\cdot)$ requires the same replacement of the functional equation by the appropriate differential equation. The derivation leads to Eq. 11.3 when a few other assumptions are made [18]. For our purposes, the most important of these other assumptions is compete positivity.

In the derivation of Eq. 11.3, it is assumed that the dynamical maps $V(t)$ are completely positive; hence, by the representation theorem,

$$V(t)\rho^S = \sum_\alpha W_\alpha(t)\rho^S W_\alpha^\dagger(t)$$

with some (time-dependent) Kraus operators W_α. Negative channels have an operator sum representation, so it is tempting to just define

$$V(t)\rho^S = \sum_\alpha \lambda_\alpha W_\alpha'(t)\rho^S W_\alpha'^\dagger(t)$$

and derive a new Lindblad equation that depends on the (possibly negative) eigenvalues of the Choi representation λ_α. Such an argument would imply that the only difference between the standard Lindblad equation and a "negative Lindblad equation" are coefficients, but such an argument would be misguided.

The problem with negative channels is much deeper than negative signs in the operator sum. The assumption of Markovian evolution, i.e.,

$$V(t_1)V(t_2) = V(t_1 + t_2), \quad t_1, t_2 \geq 0$$

assumes composability of the dynamical maps $V(t)$. Any two completely positive dynamical maps can be composed. A completely positive dynamical map will have a positivity domain which includes every possible reduced system state. This fact may also be true of a negative dynamical map, but it is not required. Two negative dynamical maps might have different (perhaps even non-overlapping) positivity domains, and it is not clear how two such maps could be composed. It might be assumed that the composition of two negative maps would simply result in a new negative map with a new restricted positivity domain that is a function of the positivity domains of the two original maps. Notice, however, that it would need to be proven that this new restricted positivity domain is non-empty. Many examples have already been given of negative channels with positivity domains that are not all of the reduced system space. Given two channels, if the first is a constant channel that sends every input state to a state that is not in the positivity domain of the second channel, then the combined, restricted positivity domain of their composition must be empty. This example points out a serious issue with defining composition over negative dynamical maps in a general way. It is not clear that negative dynamical maps have any kind of semigroup structure or that they can, in general, be described in terms of a generator in the manner shown above. It should be noted, however, that if the negative maps do form a continuous semigroup, then extending the above equations to include negative channels can be done straightforwardly [88].

The composition of negative channels is an open question. Without an answer to this question, more complicated questions about the mathematical structure of the set of all negative channels will not have very satisfactory answers. Completely positive dynamical maps have nice mathematical features beyond the Kraus representation, not the least of which is their amenability to the powerful mathematical techniques of semigroup theory. These features need to be better understood for negative channels, but until that understanding arrives, the Liouville-von Neumann equation is the best method for modeling negative channels with continuous baths.

CHAPTER 12

Proposed Experimental Demonstration of Negativity

One of the main ideas of the previous section is to illustrate *gedanken* experiments with non-zero negativity, and diagonal channels were used for the express purpose of creating examples that would be amenable to theoretical analysis. Those experiments did not involve statistical errors or systematic errors due to experimental limitations. The authors of [103] provide a nice discussion of how statistical errors in a process tomography experiment might lead to non-zero negativities, but no such considerations were given to the gedanken experiments presented in the previous sections. Hence, the negativity of those example channels cannot be a product of such experimental error.

Channels implemented in the lab, however, will have experimental error. There will be errors associated with preparing the reduced system (which would be manifested in the theory as errors in the sharp operation), implementing the reduced dynamics (which would be manifested in the theory as errors in the composite dynamics), and in implementing the tomography of the reduced system. The theoretical channel

$$\varepsilon(\vec{\tau}) = \left(U : \vec{\tau}^{\,\sharp} : U^\dagger \right)^\flat ,$$

with $\{\vec{\tau}^{\,\sharp}\}_i \in \mathcal{S}(\mathcal{H}^{SB})$, $U \in \mathcal{B}(\mathcal{H}^{SB})$ and $\{\vec{\tau}\}_i \in \mathcal{S}(\mathcal{H}^S)$ might be implemented in the lab as

$$\varepsilon(\vec{\tau})' = \left(U_\delta : \vec{\theta}^{\,\sharp} : U_\delta^\dagger \right)^\flat ,$$

where U_δ is some non-perfect implementation of the composite dynamics and $\vec{\theta}$ might be very different from the desired $\vec{\tau}$.

Example 12.1   For example, preparing the canonical tomography vector

$$\vec{\tau} = (|0\rangle\langle 0|, |+\rangle\langle +|, |+_i\rangle\langle +_i|, |1\rangle\langle 1|)$$

is difficult in certain experimental situations and can only be done probabilistically. The preparation of the state $|0\rangle\langle 0|$ might be done in such a way that the reduced system is actually prepared in the mixed state

$$\rho_0 = p|0\rangle\langle 0| + (1-p)\frac{I}{2}$$

with some $p \in [0, 1]$. This noisy preparation is actually of the form reported by [45] in their process tomography experiments. If every preparation resulted in a similar mixed state (i.e., the preparation of any of the tomography states leads to a $(1 - p)$ probability of producing the completely mixed state), then

$$\vec{\theta} = \left(\rho_0, \rho_+, \rho_{+_i}, \rho_1\right) \ .$$

Suppose the bath is (again) a single qubit and the sharp operation is the familiar

$$\vec{\tau}^\sharp = \vec{\tau} \otimes \left(H_d : \vec{\tau} : H_d^\dagger\right) \ .$$

The sharp operation is only defined on the tomography set, so the imperfect preparation state needs to be rewritten as

$$\rho_j = p|j\rangle\langle j| + \frac{(1 - p)}{2} I \ ,$$

where I is formed from the elements of the tomography set as $I = |j\rangle\langle j| + |k\rangle\langle k|$ and with $j = \{0, +, +_i, 1\}$ and k defined such that $\langle k|j\rangle = 0$. The sharp operation applied to the above noisy state yields

$$
\begin{aligned}
\rho_j^\sharp &= p\left(|j\rangle\langle j| \otimes H_d|j\rangle\langle j|H_d^\dagger\right) + \frac{(1 - p)}{2}\left(|j\rangle\langle j| \otimes H_d|j\rangle\langle j|H_d^\dagger\right) + \frac{(1 - p)}{2}\left(|k\rangle\langle k|^\sharp\right) \\
&= \frac{(p + 1)}{2}\left(|j\rangle\langle j| \otimes H_d|j\rangle\langle j|H_d^\dagger\right) + \frac{(1 - p)}{2}\left(|k\rangle\langle k|^\sharp\right) \ .
\end{aligned}
$$

From the above equation it is clear that, unless $p = 1$,

$$\vec{\theta}^\sharp \neq \vec{\tau}^\sharp$$

which implies

$$U\rho_j^\sharp U^\dagger \neq U|j\rangle\langle j|^\sharp U^\dagger \ ,$$

and the imperfectly prepared channel will be different from the desired channel (as expected).

The application of the sharp operation becomes suspect when considering "noisy" experiments. Originally, the sharp operation was introduced as a map that correctly accounts for the initial correlation of the reduced system and bath on the states actually created in the lab. The experimenter in the above example is not even sure what states are actually created in the lab. The completely mixed state might not be in the positivity domain of the sharp operation used in the channel definition. The sharp operation is, however, the experimenter's best guess for the initial correlation of the state he wants to create in the lab. As such, the experimenter must assume he can create states close to the states he uses to define his sharp operation (e.g., in the above example, the experimenter must assume $p \approx 1$). The theoretical predictions of the channel might not be testable (i.e., might not yield valid density matrices) without such an assumptions. It can be assumed, however, if it was the case that $p << 1$ in the above experiment, then the experimenter would declare his preparation procedure is be too flawed to actually conduct the tomography experiment. These issues are all part of the difficulty in comparing the theoretical predictions of the mathematical representations of channels to their physical counterparts, and such issues are precisely why many authors stress the importance of remembering the difference between measured experimental data and predicted experimental data given a specific model (e.g., [54]). For this discussion, it suffices to recognize that even if the sharp operation correctly describes the initial composite state correlation for the states the experimenter wishes to prepare, it might not do so for the states he actuals prepares, and the experimentally measured negativity might not match the predicted negativity in such a situation.

The imperfect preparation presented above is just a simple example of possible experimental error. Many other situations can lead to similar results, i.e., an implemented channel which does not follow theoretical predictions. Given these kinds of experimental errors, it can be expected that there will be differences in the negativity of the desired (i.e., theoretical) channel and the channel implemented in the lab. It is important to be able to tell the difference between theoretically predicted negativity and experimental error negativity.

Empirical evidence is required to confirm the theoretically predicted negativity, but experimental error cannot be predicted in detail (to eliminate its effects in the data post processing stage of the experiment) nor completely eliminated experimentally. Realistic preparation and measurement procedures will contain errors, and tomographic characterization of a channel will involve statistical errors associated with the data collection and processing. The correct form of the sharp operation and composite dynamics will involve educated guesswork about the bath and its interaction with the reduced system. All of these issues will make it very difficult to theoretically predict the negativity that might be measured in the lab. One possible solution is to perform experiments with a "controlled bath."

Every experiment will have a bath as defined in the introduction, but confirmation of theoretically predicted negativity will require creating an artificial bath under the control of the experimenter. Controlling the bath allows the experimenter to engineer the sharp operation and composite dynamics to verify the theoretically predicted negativity within some predefined statistical confidence. The experimenter will measure

$$\eta_{\text{measured}} = \eta_{\text{theoretical}} + \eta_{\text{error}} \ ,$$

where η_{error} is some error term due to experimental errors that can (hopefully) be made small enough to confirm

$$\eta_{\text{measured}} = \eta_{\text{theoretical}} \pm \Delta \ ,$$

where Δ is some "acceptable error level."

The proposals below are for experiments with just such "controlled baths" that can be implemented in modern quantum optics labs.

12.1 PHOTONIC ROOT-SWAP OPERATION

Quantum computing (or "quantum information processing") involves the application of a desired unitary operation to a specifically prepared initial state to achieve an output state that represents a solution to some computational problem. The desired unitary operation can be represented in terms of individual single or multiple qubit operations called "gates." Many gates are important in the quantum computing community, including the controlled NOT (CX) and the Toffoli gate, and are well studied both theoretically and experimentally. Another such gate is the two qubit "root-swap" (or $\sqrt{Sw}$) gate which is applied twice to swap the state of two qubits. The root swap gate is important to us because it leads to a negative channel for one of the input qubits.

The root swap gate is defined as

$$
U_{\sqrt{Sw}} = \frac{1}{\sqrt{2}} \begin{pmatrix} \sqrt{2} & 0 & 0 & 0 \\ 0 & 1 & i & 0 \\ 0 & i & 1 & 0 \\ 0 & 0 & 0 & \sqrt{2} \end{pmatrix} .
$$

One of the input qubits will be labeled the "reduced system" and the other will be the "bath." The sharp operation relating the initial states of these two qubits will be the familiar

$$
\vec{\tau}^{\#} = \vec{\tau} \otimes \left(H_d : \vec{\tau} : H_d^{\dagger} \right) ,
$$

where $\vec{\tau}$ is the canonical tomography vector and H_d is the Hadamard gate. The output of the channel $(\vec{\tau}, \sharp, U_{\sqrt{Sw}})$ will be found in the usual manner, i.e.,

$$
\varepsilon(\vec{\tau}) = \left(U_{\sqrt{Sw}} : \vec{\tau}^{\#} : U_{\sqrt{Sw}}^{\dagger} \right)^{\flat}
$$

which leads to the following Choi representation of the channel:

$$
C_{\sqrt{Sw}} = \mathbf{C} \odot \varepsilon(\vec{\tau}) = \begin{pmatrix} \frac{3}{4} & -\frac{i}{2\sqrt{2}} & \frac{1}{4} & \frac{\frac{1}{2}+\frac{i}{2}}{\sqrt{2}} \\ \frac{i}{2\sqrt{2}} & \frac{1}{4} & \frac{\frac{1}{2}-\frac{i}{2}}{\sqrt{2}} & -\frac{1}{4} \\ \frac{1}{4} & \frac{\frac{1}{2}+\frac{i}{2}}{\sqrt{2}} & \frac{1}{4} & -\frac{i}{2\sqrt{2}} \\ \frac{\frac{1}{2}-\frac{i}{2}}{\sqrt{2}} & -\frac{1}{4} & \frac{i}{2\sqrt{2}} & \frac{3}{4} \end{pmatrix} .
$$

Notice

$$
\mathrm{spec}(C_{\sqrt{Sw}}) = \left(\frac{1}{2}\left(1 + \sqrt{2+\sqrt{2}}\right), \frac{1}{2}\left(1 + \sqrt{2-\sqrt{2}}\right), \right.
$$
$$
\left. \frac{1}{2}\left(1 - \sqrt{2+\sqrt{2}}\right), \frac{1}{2}\left(1 - \sqrt{2-\sqrt{2}}\right) \right)
$$

and, more importantly,

$$\eta_{\sqrt{sw}} \approx 0.149 .$$

The root-swap gate has been accomplished on polarization state photonic qubits with a fidelity of about 90% [25], and those experiments may be able to be modified to experimentally test negativity calculations.

Figure 12.1 shows the optical set-up used by Cernoch et al. [25]. The yellow box in Fig.12.1

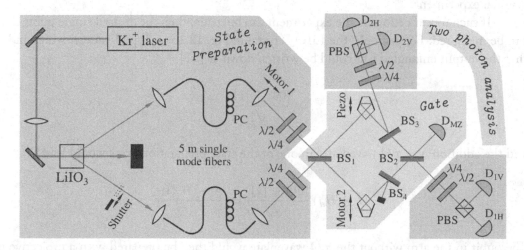

Figure 12.1: This is the optical set-up used in [25] to implement a root swap gate. This experiment can be conducted in a slightly different manner to verify the theoretically predicted negativity of 0.149 for the channel induced by this gate on one of the two input qubits. This diagram is reproduced with permission from [25].

(labeled "Gate") is the part of the process which implements the gate being investigated. The gate implementation involves the Hong-Ou-Mandel effect and arbitrary phase shifts accomplished by using the motorized prisms to change the path lengths. This optical set-up can be used to accomplish many different two qubit gates, but our interest is in the implementation of the root-swap gate. See [25] for all the details of this set-up, along with references to similar optical set-ups used to perform other gates.

The authors of the above experiment performed process tomography to demonstrate that the action of the optical set-up in Fig. 12.1 can be made very close to the desired root swap.[1] In principle, the raw data collected by this group was the superoperator representation of the two qubit channel implemented by Fig. 12.1, i.e.,

$$\mathbf{S} \odot \left(U'_{\sqrt{sw}} : \vec{\xi} : U'^{\dagger}_{\sqrt{sw}} \right)$$

---

[1]The experiment was far more than just the quantum process tomography. It also involved fidelity and entanglement measurements, but those results do not concern the discussion here.

where $U'_{\sqrt{sw}}$ is the operation actually performed by the optical set-up in Fig. 12.1 and $\vec{\xi}$ is some two qubit tomography vector. A maximum likelihood method was used on the raw data to find the "reconstructed completely positive map."

Notice that the experiment would need to be modified to calculate the negativity. The negative channel is a single qubit channel. The sharp operation on the single qubit tomography vector involves a Hadamard rotation on one of the qubits, and such a rotation was not present in the original experiment.

If such modifications to the experiment can be made, then the desired single qubit channel can be measured. For example, the $LiIO_3$ crystal in Fig. 12.1 is a Type I source, which indicates that the output entangled pair could be written down as [51]

$$\frac{|00\rangle + |11\rangle}{\sqrt{2}} \quad,$$

and the addition of a $\pi/4$ waveplate[2] in one of the state preparation arms would lead to

$$(I \otimes H_d)\frac{|00\rangle + |11\rangle}{\sqrt{2}} = \frac{|0+\rangle + |1-\rangle}{\sqrt{2}} \quad.$$

The qubit in the arm without the $\pi/4$ waveplate would then be prepared with a projective measurement (e.g., with a polarization filter). This new preparation procedure would implement the desired sharp operation, although it might significantly change the behavior of the gate.

Performing *single* qubit process tomography on one of the two qubits in this new experiment (i.e., with the correctly implemented sharp operation and root-swap gate) would lead to a superoperator (or Choi) representation from which the negativity could be measured. In the introduction, one photon was labeled the "reduced system" and the other the "bath," but the distinction only matters in the value of the negativity, not in whether or not the negativity is non-zero. This feature is important to the experimentalist who might have trouble identifying the photons in his experimental set-up. To see this idea notice,

$$\mathbf{C} \odot \left(U_{\sqrt{Sw}} : \vec{\tau}^{\sharp} : U^{\dagger}_{\sqrt{Sw}}\right)^{\flat} = \begin{pmatrix} \frac{3}{4} & -\frac{i}{2\sqrt{2}} & \frac{1}{4} & \frac{\frac{1}{2}+\frac{i}{2}}{\sqrt{2}} \\ \frac{i}{2\sqrt{2}} & \frac{1}{4} & \frac{\frac{1}{2}-\frac{i}{2}}{\sqrt{2}} & -\frac{1}{4} \\ \frac{1}{4} & \frac{\frac{1}{2}+\frac{i}{2}}{\sqrt{2}} & \frac{1}{4} & -\frac{i}{2\sqrt{2}} \\ \frac{\frac{1}{2}-\frac{i}{2}}{\sqrt{2}} & -\frac{1}{4} & \frac{i}{2\sqrt{2}} & \frac{3}{4} \end{pmatrix} \Rightarrow \eta_{\sqrt{Sw}} \approx 0.149$$

---

[2]This is a process that implements the Hadamard rotation in the sharp operation. This could be accomplished, for example, with a half-wave plate oriented at 22.5° from the optical axis [75].

and

$$\mathbf{C} \odot \mathrm{Tr}_S \left(U_{\sqrt{Sw}} : \vec{\tau}^\sharp : U^\dagger_{\sqrt{Sw}} \right) = \begin{pmatrix} \frac{3}{4} & \frac{1}{2\sqrt{2}} & \frac{1}{4} & -\frac{\frac{1}{2}+\frac{i}{2}}{\sqrt{2}} \\ \frac{1}{2\sqrt{2}} & \frac{1}{4} & \frac{\frac{1}{2}+\frac{i}{2}}{\sqrt{2}} & -\frac{1}{4} \\ \frac{1}{4} & \frac{\frac{1}{2}-\frac{i}{2}}{\sqrt{2}} & \frac{1}{4} & -\frac{1}{2\sqrt{2}} \\ -\frac{\frac{1}{2}-\frac{i}{2}}{\sqrt{2}} & -\frac{1}{4} & -\frac{1}{2\sqrt{2}} & \frac{3}{4} \end{pmatrix} \Rightarrow \eta_{\sqrt{Sw}} \approx 0.126 \; .$$

Tracing out either the "bath" or "reduced system" qubit in this experiment will lead to a negative channel. The negativity can be experimentally determined as some value $\eta'_{\sqrt{Sw}}$, and it is expected that

$$\eta'_{\sqrt{Sw}} = \eta_{\sqrt{Sw}} \pm \epsilon$$

where ϵ is some small error term due to experimental (and statistical) error. As stated before, such error is expected and cannot be eliminated completely. Error in measurement is a part of experimental physics [96]. Notice, however, unless $\epsilon \sim 10^{-1}$, the negativity found in the experiment should be greater than zero independent of which qubit is traced out.

Further confirmation of the negative channel can be found by comparing the experimentally determined superoperator representation to the theoretically expected superoperator representation. For example, a diamond norm[3] distance can be found between the two superoperators as

$$\delta = ||S_m - S_t||_\diamond$$

where S_t is the superoperator representation determined by the tomography data and

$$S_m = \mathbf{S} \odot \left(U_{\sqrt{Sw}} : \vec{\tau}^\sharp : U^\dagger_{\sqrt{Sw}} \right)^\flat$$

or

$$S_m = \mathbf{S} \odot \mathrm{Tr}_S \left(U_{\sqrt{Sw}} : \vec{\tau}^\sharp : U^\dagger_{\sqrt{Sw}} \right) \; .$$

If δ is sufficiently small (in the same way that ϵ above needs to be sufficiently small), then the measured superoperator representation S_t would be said to be the expected representation of the desired single qubit channel.

This proposed experiment is (theoretically) a straightforward extension of the Cernoch et al. [25] experiment. The experimental difficulties in implementing the suggested changes may be difficult, but they may not be insurmountable.

---

[3]The diamond norm is a popular norm for comparing quantum operations. It has "a natural operational interpretation: it measures how well one can distinguish between two transformations by applying them to a state of arbitrarily large dimension" [8] and was originally introduced in [2].

12.2 PHOTONIC CZ OPERATION

The root swap gate is not the only two qubit gate that can be used to demonstrate a negative single qubit channel. Another well-studied gate is the controlled phase (CZ) gate which is defined as

$$CZ = \begin{pmatrix} 1 & 0 & 0 & 0 \\ 0 & 1 & 0 & 0 \\ 0 & 0 & 1 & 0 \\ 0 & 0 & 0 & -1 \end{pmatrix} .$$

If one of the qubits is labeled the "reduced system" and the other the "bath," then the familiar sharp operation on the canonical tomography vector, i.e.,

$$\vec{\tau}^{\#} = \vec{\tau} \otimes \left(H_d : \vec{\tau} : H_d \right) ,$$

will lead to reduced dynamics of

$$\varepsilon(\vec{\tau}) = \left(CZ : \vec{\tau}^{\#} : CZ^{\dagger} \right)^{\flat} .$$

The labeling will, just as in the previous subsection, be unimportant to the main idea (i.e., demonstrating a negative channel) because the negativity of the single qubit channel will be non-zero independently of which qubit is traced out of the experiment. Notice,

$$\mathbf{C} \odot \left(CZ : \vec{\tau}^{\#} : CZ^{\dagger} \right)^{\flat} = \begin{pmatrix} 1 & 0 & 0 & \frac{1}{2} \\ 0 & 0 & \frac{1}{2} & 0 \\ 0 & \frac{1}{2} & 0 & 0 \\ \frac{1}{2} & 0 & 0 & 1 \end{pmatrix} \Rightarrow \eta_{CZ} \approx 0.167 \qquad (12.1)$$

and

$$\mathbf{C} \odot \mathrm{Tr}_S \left(CZ : \vec{\tau}^{\#} : CZ^{\dagger} \right) = \begin{pmatrix} \frac{1}{2} & \frac{1}{2} & \frac{1}{2} & -\frac{1}{2} - \frac{i}{2} \\ \frac{1}{2} & \frac{1}{2} & -\frac{1}{2} - \frac{i}{2} & -\frac{1}{2} \\ \frac{1}{2} & -\frac{1}{2} + \frac{i}{2} & \frac{1}{2} & \frac{1}{2} \\ -\frac{1}{2} + \frac{i}{2} & -\frac{1}{2} & \frac{1}{2} & \frac{1}{2} \end{pmatrix} \Rightarrow \eta_{CZ} \approx 0.232 .$$

These expected negativities are on the same order as those of the previous subsection.

Many different proposals exist for implementing a CZ gate with an optical set-up. A good overview can be found in [49]. Hofmann and Takeuchi suggest a way to implement a CZ gate using only beam splitters and postselection [43], and Kiesel et al. have an another (but similarly simple) design for a CZ gate using polarization dependent beamsplitters [50]. The main idea is that implementations of this gate in optical set-ups are well studied. Any of these implementations may be capable of demonstrating the proposed negative channel.

The complete experiment would involve the preparation of the tomography vector and sharp operation, and then the application of one of the above implementations of the CZ gate. The desired sharp operation is straightforward to implement in an optical set-up and can be done in exactly the same manner as described in the previous subsection: a nonlinear crystal can be used to create an entangled pair of qubits, one of which is passed through a $\pi/4$ waveplate, the other of which is prepared with polarization filters. The choice of encoding the qubit in the polarization of the photons means all of the desired operations can be accomplished with well understood polarization optics. Most of the proposals for implementing the CZ gate discussed above also use the polarization state of a photon as the qubit.

These experiments provide exactly the desired "controlled bath" situation needed to experimentally study and verify the theoretical predictions of negativity. The sharp operation can be changed by interchanging the $\pi/4$ waveplate in the preparation stage of the set-up with some other waveplate or with something more complicated like an EOM.[4] In this way, the theoretical predications concerning the impact of the sharp operation can be tested experimentally.

The next logical step in understanding negative channels would be comparing the quantum process tomography data (without any kind of complete positivity forcing post processing) to simple models of more realistic baths. The theoretical task in this process would be quite difficult because the sharp operations could only be assumed (or randomly guessed). Negative channels should still be expected in these more "realistic" experiments (i.e., experiments without controlled baths), but the theoretical analysis of these experiment suffers the same problem that any open system analysis would suffer: finding the proper form for the bath. Of course, such theoretical difficulties are irrelevant to the data collection process, and it can safely be assumed that experimental process tomography of negative channels has already happened in the literature. This topic is explored further in the next section.

---

[4]An electo-optic modulater (EOM) is a crystal placed in an electric field that can be used to modulate the polarization of incoming photons. The polarization of outgoing photons is related to the electric field strength applied to the crystal.

CHAPTER 13

Implications of Negative Channels

The negativity of complicated experiments without controlled baths can be measured, and if the negativity were found to be non-zero in experiments predicted to have non-zero negativity, then the evidence for negative channels would be convincing, even if the exact negativity of such complicated experiments could not be predicted. Such complicated experiments are actually common in the quantum information community and negative channels have probably already been observed (and promptly "corrected" in most cases).

Acceptance of the idea of negative channels might not change the field of quantum information in a fundamental sense, e.g., fundamental results such as no cloning and no signaling will still hold in general. But, several small changes will need to be made in current experimental methods and some theoretical ideas will need to be changed to incorporate negativity. This section aims to highlight a few examples of such changes and is a brief outline of some open questions about negative channels.

13.1 EXISTING EXPERIMENTAL EVIDENCE OF NEGATIVE CHANNELS

Environmental noise is considered a major obstacle to NMR systems as viable quantum information processing technologies [17]. Such strong ties between the reduced system and the bath[1] might be expected to lead to negative channels.

Consider an experiment conducted by Cory et al. [101], in which process tomography is performed on a three-qubit NMR quantum information processor. The process being investigated is the quantum Fourier transform. The process tomography of this experiment leads to a "non-completely positive superoperator," i.e., a negative channel.

The authors state that the "the spatial inhomogeneity in the RF (radio-frequency) field over the sample volume" [101] is the major contributing factoring to the observed negativity. They go on to say that the superoperator measured under such conditions "cannot be expected to precisely correspond to any physical process" [101], i.e., it cannot be expected to be completely positive.

Post-processing of the data is a way to guess what the "correct" (i.e., completely positive) process tomography data would be if the experimenters could implement the experiment without

---

[1]For most NMR systems, the reduced system would be the individual nuclear spins addressed during the experiment and the bath would be the spin bath created by all the background spins of the material.

error. To this end, the authors employ a "CP-filtering" technique [41] and a "positivity" ϱ. The positivity can be shown to be related to the negativity by the expression

$$\eta = \frac{1-\varrho}{2-\varrho} \ .$$

The "CP-filtering technique" used by the authors forces the positivity to be 1, and it is clear that such a post processing method will always lead to a channel with zero negativity. The authors take $\varrho = 1$ as their condition for completely positivity in the same manner that $\eta = 0$ if and only if the channel is completely positive, and the "CP filtering technique" used in the post processing of the experimental data guarantees this result.

The positivity of the experimental data in this NMR experiment was $\varrho = 0.60$, which corresponds to a negativity of $\eta \approx 0.29$. This negativity was unexpected by the authors, but they note "'...even though imposing the complete positivity constraint on the experimental observations did not change the supermatrix very much, the change was distinctly in the right direction since it improved the correlation with both the simulated and theoretical supermatrices." The "CP-filtering technique" is justified by better agreement of the "filtered" experimental data and the numerical simulations than between the raw experimental data and the numerical simulations.

The authors probably very much desired a preparation procedure that would be represented by a sharp operation of the form $\rho^S \otimes \tau^B$ where ρ^S is the desired state of the reduced system and τ^B is some fixed state of the bath. This sharp operation would always lead to completely positive dynamics, and in that sense, the negativity is a mistake in the implementation of the desired preparation procedure.

The point of this subsection is to illustrate an impact of accepting negative channels as physically reasonable (i.e., not "mistakes"). An experiment similar to the one in [101] could be performed with numerical simulations which predict the observed negativities. Such experiments would provide deeper insight not only into the experimental preparation procedure (and perhaps how to fix them) but also into the relationship between the reduced system and the bath.

Weinstein et al. [101] is clear, in depth, and thorough. It is also an example of something that is pervasive in process tomography experiments: post processing of experimental data to force complete positivity.

13.2 RESTRICTIONS IMPLIED BY COMPLETE POSITIVITY

An a priori requirement of a vanishing negativity puts strong restrictions on theoretical models of quantum dynamics. Perhaps most famously, Sudarshan [40] showed that complete positivity implies

$$T_1 \geq \frac{1}{2}T_2 \ , \tag{13.1}$$

where T_1 is the relaxation time for the magnetic polarization in the z direction and T_2 is the relaxation time for the magnetic polarization in the x and y directions [15, 39, 91]. T_1 is referred to as the "population" or "occupation" relaxation time, and T_2 is called the "phase" relaxation time. These terms have found application throughout the field of quantum information, but they originated in the study of nuclear magnetization in the presence of external magnetic fields. This result was derived directly from the Lindblad equation in [39], but it can de derived from Redfield theory [91] or from other common open system approximations [20, 55]. All such derivations assume complete positivity.

The relaxation times of qubits are extremely important to understand. Qubits must have the longest possible relaxations times for useful quantum information processing, and unjustified assumptions such as Eq. 13.1 can hamper the field's attempts at engineering such devices. The relationship between the occupation and phase relaxation times can be reconsidered using negative channels, but, as was pointed out in the discussion of multi-qubit baths, negative channels with continuous baths are not very amenable to current analytical techniques. Several basic questions about the mathematical structure of negative channels and the properties of their composition need to be understood before this problem can be approached rigorously. Experimental measurements are independent of such concerns, but it is important for the experimentalist to measure both times separately and not rely on Eq. 13.1 to relate them until it can be formally understand in light of negative channels. It should also be noted that complete positivity puts similar constraints on N-level channels (not just qubit channels) [83].

Quantum error correction was introduced into the field of quantum computing to answer criticisms about the viability of doing something useful with realistic (i.e., noisy, unreliable) quantum systems [1, 7, 68]. It is now widely believed that any physically realized quantum information processing device will involve copious amounts of error correction circuitry to contend with the unavoidable noise of quantum systems. The ubiquitous assumption of complete positivity can be found throughout the quantum information subfield of quantum error correction.

Shabani and Lidar [86] pointed out that most fault tolerant error correction schemes assume the entire system starts out in a product state and can be defined as a product state after every error correction step. The assumption of an initial product state might be wrong because of faulty preparation procedures, but the assumption of a product state after every error correction would only be true if the error correction step was "perfect." As Shabani and Lidar stated "However, FT-QEC [fault tolerant quantum error correction] allows for the fact that the error correction step is almost never perfect, which means that there is a residual correlation between system and bath at t_1 [the time step at which the "instantaneous error correction" procedure takes place]." [86] This "residual correlation" after the error correction steps implies that the composite system state cannot be described as a product state and the single qubit channel under investigation cannot be assumed to be completely positive. For a detailed discussion of these issues see [7].

13.3 EXPANDING THE DEFINITION OF "PHYSICALLY REASONABLE"

Enumerating all the possible fallouts from incorporating negative channels into the current theory of quantum information is not the goal here. For example, it has recently been argued that negative channels can violate the Holevo bound [63], which is an interesting and important result that was not addressed above. Negative maps can also be used to improve distinguishability [24], which is not possible with completely positive maps.

The idea of this section is two fold. First, acceptance of negative channels will involve modifications to some core ideas in the field (like the Choi-Jamiolkowski isomorphism and the Holevo bound). Such changes may or may not be straightforward, but negative channels are physically reasonable[2] and must be included in a complete theory of quantum information. Second, it should be noticed that the examples presented here are (for the most part) not new issues. Havel et al. noticed negative channels in their experimental data in 2004 [41], Shabani has addressed the problem of negative channels in quantum error correction in 2009, and Sudarshen et al. addressed the T_2/T_1 issue in 1976 and 1978. Negative channels have been theoretically recognized for a while, but have been ignored for reasons that are probably closer to indifference rather than resistance.

The issue of negativity was not something that concerned a lot of physicists until quantum information became prevalent. The expansion of quantum information theory into quantum information experiments began to force theorist and experimentalist alike to consider "noisy" quantum systems. This newfound interest in open quantum systems has lead to a renewed interest in negative channels, including their causes and implications. This section is meant merely to illustrate that negative channels will lead to some changes in the current theory of quantum information and these changes are not completely unexpected.

---

[2]This phrase has been used extensively. Specifically, "physically reasonable" is being used to mean "can exist in nature," "is consistent with theoretical quantum mechanics," "can be verified experimentally," and "can be understood using established theoretical techniques." In this case, those "established theoretical techniques" are the techniques of open quantum systems.

CHAPTER 14

Uses for Negative Channels

So far, one of the most important questions about negativity has been completely ignored in this work. Is negativity useful? This section will focus on that question.

14.1 PROBING THE BATH

14.1.1 COUPLING ALONE

Consider a two qubit composite system with composite dynamics defined as

$$
U_\theta = \begin{pmatrix} 1 & 0 & 0 & 0 \\ 0 & \cos\theta & \sin\theta & 0 \\ 0 & -\sin\theta & \cos\theta & 0 \\ 0 & 0 & 0 & 1 \end{pmatrix} .
$$

If the sharp operation takes the form

$$
\vec{\tau}^{\#} = \vec{\tau} \otimes \left(H_d : \vec{\tau} : H_d^{\dagger} \right) ,
$$

where H_d is the Hadamard operation, and is defined on the canonical tomography vector $\vec{\tau}$, then the Choi representation of a single qubit channel is

$$
\begin{aligned}
C_\theta &= \mathbf{C} \odot \left(U_\theta : \vec{\tau}^{\#} : U_\theta^{\dagger} \right)^{\flat} \\
&= \begin{pmatrix} A & B \\ B^{\dagger} & C \end{pmatrix} ,
\end{aligned}
$$

where

$$
A = \begin{pmatrix} \frac{1}{4}(3 + \cos(2\theta)) & -\frac{\sin\theta}{2} \\ -\frac{\sin\theta}{2} & \frac{\sin^2\theta}{2} \end{pmatrix} ,
$$

$$
B = \begin{pmatrix} \frac{1}{2}\sin\theta(-i\cos\theta + \sin\theta) & \cos\theta + \left(\frac{1}{2} + \frac{i}{2}\right)\sin\theta \\ \left(\frac{1}{2} + \frac{i}{2}\right)\sin\theta & \frac{1}{4}\left(-1 + e^{2i\theta}\right) \end{pmatrix} ,
$$

and

$$
C = \begin{pmatrix} \frac{\sin^2\theta}{2} & -\frac{\sin\theta}{2} \\ \frac{1}{4}\left(-1 + e^{-2i\theta}\right) & -\frac{\sin\theta}{2} & \frac{1}{4}(3 + \cos(2\theta)) \end{pmatrix} .
$$

Notice $\theta = 0$ and $\theta = 2\pi$ leads to $U_\theta = I$ where I is the two qubit identity operator, and $\theta = \pi$ leads to $U_\theta = \sigma_3 \otimes \sigma_3$. Both of these composite dynamics are in local unitary form, so these three angles lead to a vanishing negativity. Notice that U_θ is cyclic in the sense that it will be in local unitary form (and, therefore, lead to a vanishing negativity) if $\theta = n\pi$ for $n \in \mathbb{Z}$. Define, η_θ to be the negativity of the channel represented by C_θ. The negativity can be plotted as function of θ (see Fig. 14.1) to reveal a maximum negativity of $\eta_\theta \approx 0.24$.

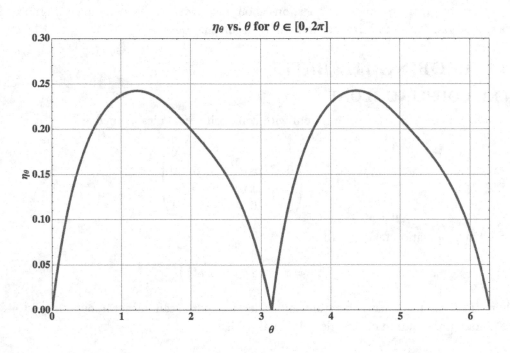

Figure 14.1: The negativity η_θ can be plotted as a function of θ to show the dependency of the negativity on U_θ. See the text for definitions of U_θ and C_θ. The points of vanishing negativity and the periodicity of this plot are also discussed in the text.

The negativity η_θ is a function of θ and can, in principle, be used to gain information about θ. The negativity can be measured and if the above theoretical definition of the channel is assumed to be true, then θ can simply be read off Fig. 14.1. The negativity η_θ is measured in a single qubit tomography experiment, but U_θ cannot be directly measured in any such experiment because the composite system contains a qubit defined to be beyond the reach of the experimenter (i.e., the bath qubit).

14.1.2 CORRELATION ALONE

Consider a similar, but different example. The new composite dynamics are defined by the controlled phase gate, i.e.,

$$CZ = \begin{pmatrix} 1 & 0 & 0 & 0 \\ 0 & 1 & 0 & 0 \\ 0 & 0 & 1 & 0 \\ 0 & 0 & 0 & -1 \end{pmatrix}$$

and the sharp operation is defined on the canonical tomography vector $\vec{\tau}$ as

$$\vec{\tau}^{\#\alpha} = \vec{\tau} \otimes \left(U_\alpha : \vec{\tau} : U_\alpha^\dagger \right)$$

with

$$U_\alpha = \alpha \sigma_1 + \sqrt{(1-\alpha^2)} \sigma_3 \ ,$$

where $\vec{\sigma} = (\sigma_0, \sigma_1, \sigma_2, \sigma_3)$ is the standard Pauli vector discussed in the tomography section (Sec. 2). Notice, U_α is unitary if $\alpha \in [0,1]$ with $U_\alpha = H_d$ if $\alpha = 2^{-1/2}$. The Choi representation of a single qubit channel in this two qubit composite system would be

$$C_\alpha = \mathbf{C} \odot \left(CZ : \vec{\tau}^{\#\alpha} : CZ^\dagger \right)^{\flat}$$
$$= \begin{pmatrix} 1 & 0 & 0 & \alpha\sqrt{1-\alpha^2} \\ 0 & 0 & \alpha\sqrt{1-\alpha^2} & 0 \\ 0 & \alpha\sqrt{1-\alpha^2} & 0 & 0 \\ \alpha\sqrt{1-\alpha^2} & 0 & 0 & 1 \end{pmatrix} .$$

The spectrum of C_α can be written down immediately as

$$\text{spec}(C_\alpha) = \{1 - x_\alpha, -x_\alpha, x_\alpha, 1 + x_\alpha\}$$

where $x_\alpha = \alpha\sqrt{1-\alpha^2}$. The negativity of this channel η_α is bounded by

$$\alpha \in [0,1] \Rightarrow \eta_\alpha \in \left[0, \frac{1}{6}\right] \ ,$$

with $\eta_\alpha = 0$ if $\alpha = 0$ or $\alpha = 1$ and $\eta_\alpha = 1/6$ if $\alpha = 2^{-1/2}$. The negativity η_α was already calculated for the case when $U_\alpha = H_d$ (i.e., $\alpha = 2^{-1/2}$) in the discussion of proposed experiments to measure negativity (see Sec. 12.2).

The dependence of η_α on α can plotted to illustrate this idea a little more clearly (see Fig. 14.2).

Again, the negativity η_α can be measured and if the above theoretical definition of the channel is assumed to be true, then α can simply be read off Fig. 14.2. In this example, as in the previous one, measurement of the negativity in a tomography experiment grants the experimenter

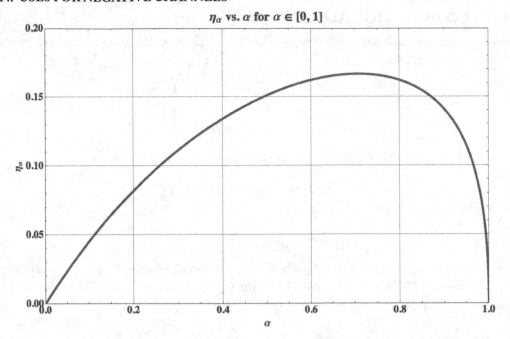

Figure 14.2: The negativity η_α can be plotted as a function of α to show the dependency of the negativity on U_α. This example, like the example plotted in Fig. 14.1, illustrates how the negativity yields information about parameters in the channel definition.

knowledge about channel parameters that cannot be measured directly. It has been argued that plotting the spectrum of the Choi representation yields information similar to the negativity [76], which is true, but the spectrum can become large for multi-qubit channels and the relationship between the eigenvalues (e.g., the trace condition of the Choi representation) makes some of the information in the spectrum redundant. The negativity is meant to be a measure of the system-bath coupling and correlation that condenses information from a tomography experiment into a single parameter, allowing it to be useful (and manageable) beyond the single qubit examples given here.

The above examples are artificial in the sense of comparing the experimentally measured negativity to some known analytical definition of the channel (i.e., C_θ and C_α). Typically, the experimenter will not have very detailed expectations about the form of the channel in the tomography experiment. Some assumptions might be made about the form of the sharp operation or composite dynamics, but it is rare to have a model of the channel complete enough (or which the experimenter has enough confidence in) to do the type of direct comparison between theory and experiment described in the examples. Typically, the experimenter would be doing tomography experiments precisely to figure out which assumptions about the sharp and composite dynamics are reasonable. Nevertheless, even without precise, confidence-worthy models of the experimental channels, measurement of the negativity will provide information about the composite system. A good example of this point is the experiment conducted in [101], which has already been discussed in Sec. 13.1, where the authors used the negativity (which they called the "positivity") to try to determine possible problems with their experimental setup.

14.1.3 COUPLING AND CORRELATION TOGETHER

Consider a slightly more complicated example with the composite dynamics given by U_θ and the sharp operation from the above example, i.e., consider the channel

$$\varepsilon(\vec{\tau}) = \left(U_\theta : \vec{\tau}^{\sharp\alpha} : U_\theta^\dagger\right)^\flat .$$

This single qubit channel combines the two above examples and will yield a negativity dependent on both the "correlation" (i.e., α) and the "coupling" (i.e., θ). The Choi representation of this channel is

$$\begin{aligned} C_{\theta\alpha} &= \mathbf{C} \odot \varepsilon(\vec{\tau}) \\ &= \begin{pmatrix} A & B \\ B^\dagger & C \end{pmatrix} , \end{aligned}$$

where

$$A = \begin{pmatrix} 1 - \alpha^2 + \alpha^2 \cos^2\theta & -\alpha\sqrt{1-\alpha^2}\sin\theta \\ -\alpha\sqrt{1-\alpha^2}\sin\theta & \alpha^2\sin^2\theta \end{pmatrix}$$

$$B = \begin{pmatrix} c_1 & c_2 \\ (1+i)\alpha\sqrt{1-\alpha^2}\sin\theta & c_3 \end{pmatrix} ,$$

and

$$C = \begin{pmatrix} \alpha^2\sin^2\theta & -\alpha\sqrt{1-\alpha^2}\sin\theta \\ -\alpha\sqrt{1-\alpha^2}\sin\theta & 1 - \alpha^2 + \alpha^2\cos^2\theta \end{pmatrix}$$

with

$$\begin{aligned} c_1 &= \frac{1}{2}\sin\theta\left(\left((-1-i)+2\alpha^2\right)\cos\theta + 2\alpha\sqrt{1-\alpha^2}\sin\theta\right) , \\ c_2 &= \cos\theta + (1+i)\alpha\sqrt{1-\alpha^2}\sin\theta , \end{aligned}$$

and

$$c_3 = \frac{1}{2}\left((1+i)\cos\theta\sin\theta - 2\alpha^2\cos\theta\sin\theta - 2\alpha\sqrt{1-\alpha^2}\sin^2\theta\right) .$$

Notice $\theta = 0$ and $\theta = 2\pi$ yield

$$C_{0\alpha} = C_{2\pi\alpha} = \begin{pmatrix} 1 & 0 & 0 & 1 \\ 0 & 0 & 0 & 0 \\ 0 & 0 & 0 & 0 \\ 1 & 0 & 0 & 1 \end{pmatrix}$$

and $\theta = \pi$ yields

$$C_{\pi\alpha} = \begin{pmatrix} 1 & 0 & 0 & -1 \\ 0 & 0 & 0 & 0 \\ 0 & 0 & 0 & 0 \\ -1 & 0 & 0 & 1 \end{pmatrix}.$$

All of these Choi representations $C_{0\alpha}$, $C_{2\pi\alpha}$, and $C_{\pi\alpha}$ represent channels with vanishing negativities independent of the value of α. Notice also that the periodicity of U_θ is still expected to lead to periodicity in the negativity for this example. The composite dynamics cyclically have a local unitary form (as was explained in the first example), and local unitary composite dynamics lead to a vanishing negativity for any sharp operation, i.e., independent of α. So, there will be periodicity about the point $\theta = \pi$ in this example just an there was in the first example.

The negativity of the channel represented by $C_{\theta\alpha}$ can be plotted as a function of the full two parameter space as a contour map (see Fig. 14.4). The negativity can also be plotted as a surface in this two dimensional parameter space, which is done in Fig. 14.3 for a single period of θ (i.e., for $\theta \in [0, \pi]$).

The measured value $\eta_{\theta\alpha}$ cannot uniquely identify a location in the two dimensional parameter space plotted in Fig. 14.4, and this inability is precisely the frustrating limitation of the bath information hidden in the negativity value.

If the experimenter were able to measure both the negativity and the initial system-bath correlation, then the experimenter would still not be able to draw any conclusions about the causal relationship between the two because the negativity would still be influenced by the unknown coupling. If he were able to measure the negativity and the coupling, then the system-bath correlation would act in the same manner. In most situations, the experimenter will only be able to measure the negativity and will be unable to understand the causal relationship between those measurements and the preparation procedure (i.e., the correlation) or the composite dynamics (i.e., the coupling), unless he is able to control for the relationship between them.

Notice that the measured negativity will limit the possible values of θ and α to some subset of the total parameter space which may be substantially smaller and might be helpful to the experimenter. For example, if the experimenter is attempting to use the measured negativity to develop an empirical model of the coupling and correlation, then this smaller parameter space may make the task of comparing numerical simulations to measured data easier.

14.1.4 DETERMINING THE COMPLETELY POSITIVE PARAMETER SPACE

Consider a "controlled bath" type experiment implemented in the lab using the polarization of two maximally entangled photons as qubits. One photon would act as the reduced system and the other as the bath. It would be possible to implement the sharp operation presented in Sec. 14.1.1 as a projective measurement on the reduced system photon after applying a rotation to the bath qubit photon. The negativity could then be plotted as a function of time. Repeating this experiment multiple times with different initial rotation angles for the bath qubit would indicate to the

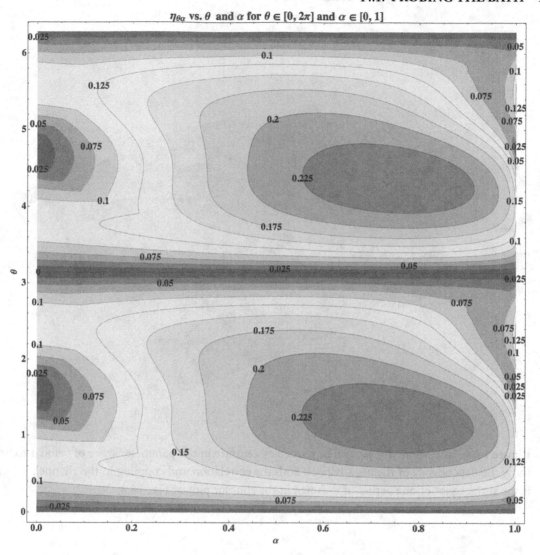

Figure 14.3: The negativity $\eta_{\theta\alpha}$ can be plotted as a function of θ and α to show the dependency of the negativity on both the correlation and coupling in the channel. The contours are labeled with the values of the negativity $\eta_{\theta\alpha}$. The maximum negativity is achieved in the red area of the plot and the minimum is achieved in the blue area of the plot. See the text for precise definitions of "correlation" and "coupling."

experimenter when the composite dynamics are described by local unitaries. As such, empirically determining when the negativity is zero for a large number of different initial system-bath corre-

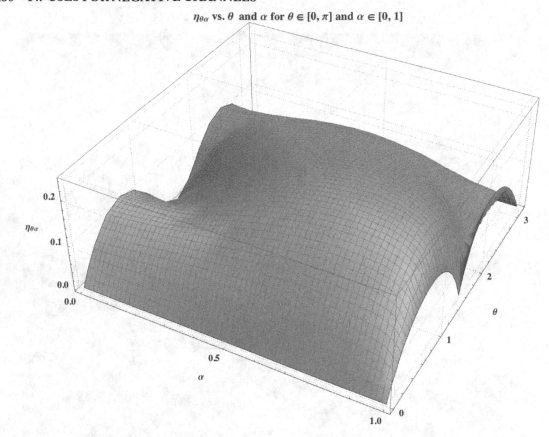

$\eta_{\theta\alpha}$ vs. θ and α for $\theta \in [0, \pi]$ and $\alpha \in [0, 1]$

Figure 14.4: The negativity $\eta_{\theta\alpha}$ can be plotted as a surface in the parameter space of θ and α to better show the dependency of the negativity on both the correlation and coupling in the channel. This plot is the surface plot of one period of θ shown in the contour plot of Fig. 14.4.

lations will allow the experimenter to be reasonably confident of when the composite dynamics are described by local unitaries. It follows that an experiment that measures the changes in negativity over time for several different initial rotation angles of the bath (i.e., ϕ) can be used to determine when the composite dynamics are in local unitary form without ever knowing anything about the bath dynamics directly. Such "controlled bath" experiments could also be used to understand preparation procedures.

These are specific examples of the most straightforward usefulness of the negativity, mapping the completely positive parameter space. For example, many of the example channels already shown have a Choi representation that takes the form

$$
\mathbf{C} = \begin{pmatrix} 1 & 0 & 0 & x \\ 0 & 0 & y & 0 \\ 0 & y^* & 0 & 0 \\ x^* & 0 & 0 & 1 \end{pmatrix} .
$$

The spectrum of $\mathbf{C}$ can be written down as

$$
\text{spec}(\mathbf{C}) = \left(1 - \sqrt{xx^*}, 1 + \sqrt{xx^*}, -\sqrt{yy^*}, \sqrt{yy^*} \right) .
$$

Notice that $yy^* = 0$ and $xx^* \leq 1$ are sufficient conditions for this channel to have a vanishing negativity. These conditions will only be met at specific points in the parameter space of the experiment. This idea can be illustrated by plotting a few points of vanishing negativity in the 3-dimensional parameter space of (k_z, t, ϕ) for the Rabi channel of Sec. 9.2 (see Fig. 14.5). The assumption of a fixed (Hadamard) rotation in the sharp operation of Sec. 9.2 has been dropped to produce this plot. Also notice that the planes $t = 0$ and $k_z = 0$ are not plotted because $t = 0$ leads to trivial composite dynamics and $k_z = 0$ leads to local unitary composite dynamics. Both situations imply complete positivity and would clutter the plot unnecessarily. As such, the plot is over the range $\{k_z, t, \phi\} \in (0, 2\pi]$.

14.2 SPECULATION ON UTILITIES

The "controlled bath" situation described in the previous subsection allows the experimenter to circumvent the correlation-coupling delineation problem in the negativity measurement. If the correlation is precisely controlled, then the negativity can be related directly to the coupling. The inverse is also true: if an experimenter precisely knows the composite dynamics implemented in a "controlled bath" experiment, then the negativity can be related directly to the correlation. Such an experiment might be used to study preparation procedures or measure correlations in multi-qubit initial states.

The utility of the negativity is apparent in "controlled bath" experiments because these situations are precisely the situations which lend the negativity a straightforward relationship with the bath. The negativity, however, may be useful beyond these experiments. The negativity provides information about the bath. The bath is defined by the experimenter's ignorance. Hence, the negativity provides information about part of the quantum system traditionally considered unknowable. The inscrutable effects of the bath through the coupling and correlation can be observed through negativity measurements. Such information is academically interesting and might have practical use in engineering quantum technologies and/or understanding the limitations of those technologies. A channel with a non-zero negativity has some correlation or coupling (or both) with the bath and such interactions with the bath might lead to new understanding of the information theoretic properties of quantum channels. For example, such channels might have

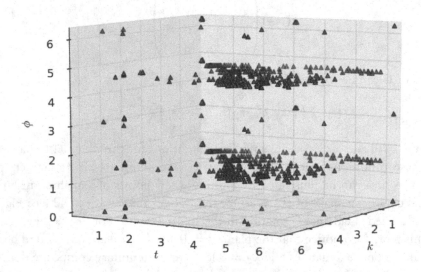

Figure 14.5: The points where the negativity of the channel described in Sec. 9 are zero in the parameter space of time t, coupling constant k_z, and the initial rotation angle of the bath qubit ϕ. This plot is meant to illustrate the idea of mapping out a completely positive parameter space.

different capacities than their completely positive counter parts, or the security of such channels might not be understood in the same way.

CHAPTER 15

Conclusions

Complete positivity is not always a physical requirement of quantum operations, and not completely positive (i.e., negative) channels need to be studied to better understand how modern quantum information theory concepts can be implemented and tested in the lab. A complete theory of quantum information must necessarily include an understanding of negative channels. Negative channels can be created in the lab, and the degree to which a channel is not completely positive (i.e., the negativity) can be measured in quantum process tomography experiments. All of these points have been made, but it should be emphasized that the dropping of the complete positivity requirement for quantum channels is somewhat expected within the community. Many different avenues of research are pointing to the abandonment of complete positivity as a physical requirement for quantum operations.

15.1 COMPLETE POSITIVITY IN RECENT YEARS

In 2007 Alicki and Lendi argued [5]:

"Complete positivity (CP) has experienced its absolute breakthrough in the wide and very active fields of quantum information, particularly in quantum computing. It has been recognized that the CP requirement must unavoidably be imposed on operations affecting only one component of entangled systems, since otherwise artificial and unphysical correlations or ill-defined states may emerge."

This quote refers to work of Sewell [85] and Alicki et al. [4] which follow the total domain argument[1] for complete positivity as a physical requirement. This remark comes from a paper entitled "Recent Developments" in [5]. Sudarshen et al. pointed out the problems of complete positivity assumptions repeatedly [39, 40, 78, 89], even pointing out the restrictions it places on the T_1 and T_2 decay rates as far back as 1978. Yet, as this recent quote shows, negative channels are not very popular in at least part of the community. The community as a whole, however, appears to be slowly embracing the idea of negative channels.

In 2006, Terno delivered a talk entitled "Non-Completely Positive Maps in Physics" [97] in which he cites the work of Štelmachovič and Bužek [92] and Życzkowski and Bengtsson. The former do not comment on complete positivity as a physical requirement, but the latter defend their work on non-completely positive maps by arguing: "Even though positive, but not completely positive, maps cannot be realized in the laboratory, they are of a great theoretical importance,..." [108]. This quote seems to imply that the authors believe negative channels are

[1]See Sec. 5.

not physical, but important nonetheless. Terno also cites Sudarshen et al. [48] who states, rather straightforwardly, "In the light of understanding gained here, it is easy to see the errors in arguments that a map describing the evolution of an open quantum system *has to be completely positive*."[2]

In 2009, Rodriguez-Rosario wrote a thesis in which he introduces an example of a negative channel and discusses the lack of complete positivity in light of non-Markovian dynamics [76]. That same year, Wood wrote a thesis titled "Non-completely positive maps: properties and applications" [103] in which he discusses the experimental evidence of negative channels seen by Havel et al. (see Sec. 13.1) and whether or not such observations can be explained as statistical errors in the tomography process. This work includes a discussion of methods with which "...one can distinguish between the distributions of CP [completely positive] and true non-CP processes with a high degree of accuracy" where the author is concerned about statistical errors in the tomography process leading to observations of a negative channel that should theoretically be completely positive. But, notice that this quote assumes the existence of "true non-CP" quantum processes. The thesis of Shabani [7], also written in 2009, extends the concepts of fault tolerant quantum computing to non-completely positive maps, explicitly pointing out that the product state argument[3] is not applicable in most quantum error correction scenarios.

The push to experimentally realize the ideas of quantum information theory has led to some direct conflicts with complete positivity. Negative channels are commonly observed in quantum process tomography experiments and a few authors (like the ones mentioned above) want to study why this happens. The theoretical study of quantum information concepts in "realistic" scenarios leads to conclusions which can only be addressed with a theory beyond completely positive maps (as pointed out by Shabani). The limitations of completely positive maps become more pronounced as quantum information theory seeks to explain more and more of the natural world. For example, consider this quote from Fleming and Hu in 2011 [36]:

"In summary, completely-positive maps are much less useful outside of the Markovian regime, as one rarely has the all-time maps. 'All-time' here meaning that the maps must describe all times wherein there may not be any correlation to anything. More precisely, such a map would have to describe the entire universe from its very birth. Typically and empirically, one usually has information only pertaining to the two-time maps of some limited set of states. These maps are not completely positive in the non-Markovian regime and not much is known about them."

Here the authors are describing the limitations of maps with vanishing negativity in a discussion of the Markov assumption in open quantum systems. Complete positivity describes an idealized theoretical model of a quantum operation useful in its mathematical simplicity but limited in its experimental applicability.

---

[2]Emphasis in original.
[3]See Sec. 5.

15.2   CLOSING REMARKS

Our motivation is nicely summed up in a quote by Yuen in his discussion of the uncertainty principle [104]:

"It seems to me that current quantum information science and technology also suffers from a number of inadequacies in its foundation. In particular, many mathematical models that have been extensively analyzed are not sufficiently connected to physical and conceptual considerations on realistic experimental situations that would allow one to draw useful conclusions for real applications . . . . It is important to remember that physicists and engineers need quantitative theories, not just qualitative or asymptotic ones, to build real systems."

A point needs to be belabored here about the study of negative channels. Complete positivity is often considered a "special case" even by proponents of negative channels. Extending quantum information theory to include negative channels is often seen as a generalization of the theory. Completely positive channels are a "special case" of negative channels, but this language should be taken only to mean that they are a very specific case. It should not be taken to mean that negative channels are not as common as completely positive channels, or that most physical situations that arise in the lab are adequately described with completely positive maps. This attitude is precisely what we were referencing when we mentioned in the previous chapter that the lack of acceptance of negative channels is probably due more to indifference than anything else.

The complete positivity assumption is not always physically reasonable. This point was made clear in the Rabi channel examples (see Sec. 9), where it was shown that the complete positivity requirement is only applicable to a very specific set of parameter values for the reduced system. An a priori assumption of complete positivity is an a priori assumption that the physical parameters of the channel have very specific values. We have shown that the energy eigenbasis evolution of a qubit channel is almost always negative if the composite Hamiltonian is time independent and the bath consists of a single qubit that is initially prepared as a Hadamard rotation of the initial state of the reduced system. This result implies that if completely positivity is required, then most mathematical constructions of such channels are not physically reasonable. Perhaps the strangest conclusion of this line of logic is that given a fixed set of parameters in the Hamiltonian, these channels are only "physical" at certain, specific points in time.

There are, of course, many situations when the complete positivity assumption is reasonable. For example, if the initial correlation is assumed to be a product state with a fixed bath, or if the composite dynamics are assumed to have local unitary form, then the reduced dynamics can be assumed to be completely positive. It should be noted, however, that any conclusions drawn from such assumptions are, as always, limited by those assumptions.

We have also shown that negative channels can be realized in the lab, and we have suggested "controlled bath" experiments that would allow models of negative channels to be verified. We have argued that experimental observations of negative channels have already occurred, but performing the suggested "controlled bath" experiments would serve as a very strong argument against complete positivity as a physical requirement. If negative channels can be observed in the lab, then complete positivity cannot be a requirement of all physical channels.

The "entanglement witness" argument is a popular argument in favor of complete positivity. For example, see [74, 80]. The argument relies on a world view that can be depicted as

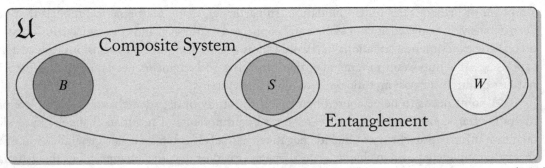

$$(15.1)$$

where $\mathfrak{U}$ is the universe, B is the bath, S is the reduced system, and W is the entanglement witness. The entanglement witness is part of neither the reduced system nor the bath, but it is still expected that the system consisting of both S and W (the system "SW") must be represented by a valid density matrix. Why is W not a part of B? It might be claimed that W is not intended to be a physical object. Instead, it might be argued, W should just be considered a "witness" to the physics. Notice, however, that the very claim of complete positivity relies on W being a physical object.

The requirement of complete positivity is a requirement of a valid (specifically positive) density matrix representation for SW. If W is not physical, why should SW be expected to have a valid density matrix representation? At most, it might be expected that $(SW)^\flat = S$, but notice that this is a requirement of consistency, not complete positivity. The system W has no direct correlation with the bath, but it imposes restrictions on the reduced dynamics because, as is often argued, it can never be known if the system W exists or not and S must always evolve into another density matrix. It has already been pointed out that negative channels can have positivity domains that cover all of the reduced system space. So, complete positivity is not required to insure the positivity of S.

Notice also that if B were maximally entangled with S, then the evolution of S would be unaffected only if the composite dynamics were in local unitary form. This statement would be true independently of the experimenter's knowledge of the existence of B. It is implied that the reduced dynamics must be completely positive because the existence of W cannot influence the channel. The existence of B, however, must influence the channel. Such logic is confusing and appears contrived. Complete positivity is a nice property, but philosophical arguments about entanglement witnesses do not provide the sound physical motivation required to impose it on all channels as an a priori assumption. These points were argued at the beginning of Ch. 2, but hopefully returning to them in this conclusion, after the introduction and analysis of several different negative channel examples, makes the point a little clearer.

As a final note, it should be recognized that most of the discussion we have presented has implications far beyond quantum information theory. Everything here was presented in the framework of quantum information theory but the conclusions extend to any open quantum system. Complete positivity is an assumption in the field of open quantum systems and that field has been applied to everything from non-equilibrium thermodynamics to high energy physics [18]. For example, the time inversion operator is often considered to not be physical because it is known to not be completely positive [21].

The proposed experimental measurements of negativity we presented are not the first proposed experimental tests of complete positivity. In the late 1990s, it was pointed out that complete positivity puts bounds on experimental parameters in both neutral kaon experiments [9, 10, 12] and neutron interferometry experiments [13]. Experiments with neutral kaon systems were proposed as a way to test complete positivity [11]. The experiments we propose are much simpler (both theoretically and experimentally) and involve the concept of negativity (i.e., quantification of the lack of complete positivity), but the proposed neutral kaon system experiment illustrates the far reach of the complete positivity assumption. Complete positivity has observable effects, and, if required, it limits the physical processes that are possible. This is a limitation felt in all of theoretical physics, not just quantum information.

There are many open questions in the study of negative channels. Our intent is to point out that the assumption of complete positivity is not always reasonable. Once these ideas are confirmed by experiment (e.g., as proposed in Sec. 12), then the study of negative channels can be used to better understand how to implement many of the currently proposed quantum technologies, and it might even lead to new proposed technologies considered impossible under the assumption of complete positivity.

APPENDIX A

Discussions

A.1 DISCUSSION OF THE REDUCED SYSTEM DEFINITION

Some subtleties of the reduced system definition can be better understood with an example. Consider two 2-level systems and a single particle shared between them (e.g., a single electron in a double quantum dot where each dot only contains two energy levels). This system can be seen schematically in Fig. A.1

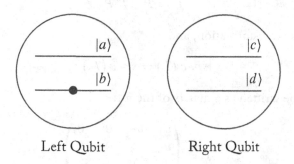

Left Qubit   Right Qubit

Figure A.1: The two-level systems are spatially separated and share a single electron between them. The electron is drawn in the lower level of the left qubit for illustrative purposes.

This example system has four energy levels, so it might feel natural to represent the system with a four dimensional Hilbert space $\mathcal{H}^T = \mathcal{H}^L \otimes \mathcal{H}^R$ where $\mathcal{H}^R$ is the two-dimensional Hilbert space associated to the right dot and $\mathcal{H}^L$ is the two-dimensional Hilbert space associated to the left dot. The above definition of the reduced system implies that the right qubit can only be defined as the "reduced system" if it is possible to "access" the system with a composite operation of the form

$$T = I \otimes U \ ,$$

where I is the identity operator on the left qubit and U is some unitary operator on the right qubit. Mathematically, the reduced system would be defined by the partial trace over $\mathcal{H}^L$.

Notice that if the four energy levels in the system are labeled $|a\rangle$, $|b\rangle$, $|c\rangle$, and $|d\rangle$ as seen in the figure, then composite dynamics T' that leave the left qubit unaffected and apply U' to the right qubit are written in that basis as a matrix of the form

$$T' = \begin{pmatrix} 1 & 0 & 0 & 0 \\ 0 & 1 & 0 & 0 \\ 0 & 0 & u'_1 & u'_2 \\ 0 & 0 & u'_3 & u'_4 \end{pmatrix}$$

where

$$U' = \begin{pmatrix} u'_1 & u'_2 \\ u'_3 & u'_4 \end{pmatrix} .$$

If a "reduced system" exists in this system that meets the definition presented here, then there must be a basis in which T' can be written in the form T, i.e., T must be similar to T' (written as $T \sim T'$). Similarity implies

$$T = P^{-1} T' P ,$$

where P is some basis transformation, and

$$spec\,(T) = spec\,(T') .$$

The operator T can be written as a matrix of the form

$$T = \begin{pmatrix} u_1 & u_2 & 0 & 0 \\ u_3 & u_4 & 0 & 0 \\ 0 & 0 & u_1 & u_2 \\ 0 & 0 & u_3 & u_4 \end{pmatrix}$$

where

$$U = \begin{pmatrix} u_1 & u_2 \\ u_3 & u_4 \end{pmatrix} .$$

Notice

$$spec\,(T) = \{\lambda_+, \lambda_+, \lambda_-, \lambda_-\}$$

where

$$spec\,(U) = \{\lambda_+, \lambda_-\} ,$$

and

$$spec\,(T') = \{1, 1, \lambda'_+, \lambda'_-\}$$

where

$$spec\,(U') = \{\lambda'_+, \lambda'_-\} .$$

As such,

$$T \sim T' \rightarrow \lambda'_+ = \lambda'_- \equiv \lambda' \rightarrow U' = \lambda' I ,$$

i.e., if T and T' are similar, then U' must be some multiple of the identity matrix. Global phases are irrelevant in quantum mechanics, hence a multiple of the identity matrix will act the same as the identity matrix on the right qubit. The conclusion seems to be that the given definition of the "reduced system" implies that the reduced system can only be formally defined for this system if the composite dynamics are trivial.

While such a conclusion may be unsatisfying, it points out the need to take care in defining the mathematical structure of the system under investigation. If the experimenter can only access the right qubit, then the mathematical model of the composite system must reflect that in a complete way. The desire in the construction of T was "apply U to the right qubit and do nothing to the left qubit," but a unitary U that is restricted to the subspace spanned by $\{|c\rangle, |d\rangle\}$ does not consider the fact that there may be no electron in the right qubit at all. In the mathematical model of the system presented above, if the "reduced system" was defined as the right qubit, then it would be possible to have a "reduced system" that may not contain the electron. The "state" of the system is determined by which energy level is occupied by the electron, hence the "reduced system" could be undefined (because the only electron in the composite system may be in the "bath").

It may be the case that the experimenter is only able to access the right qubit for physical reasons. The above reasoning shows that this physical situation is not described well by a four-dimensional Hilbert space for the composite system, but that does not mean that a reduced system cannot be mathematically defined for the system. The composite Hilbert can still be defined as $\mathcal{H}^T = \mathcal{H}^L \otimes \mathcal{H}^R$, but the subsystem Hilbert spaces $\mathcal{H}^R$ and $\mathcal{H}^L$ could be 3-dimensional (which implies $\mathcal{H}^T$ would be 9-dimensional). The addition of a third (or "vacuum") energy level would account for the possibility of zero electrons in the reduced system. The reduced system would still be defined by "tracing out" the Hilbert space associated to the left qubit $\mathcal{H}^L$. There are many such ways to mathematically model a "physical" definition of the reduced system being just the right qubit in this system. It is important to recognize, however, that the given definition for "reduced system" may require a mathematical model of the composite system that is not the most "natural" (or "obvious") choice.

A.2   GENERAL REDUCED SYSTEM DYNAMICS

The initial composite state has an eigendecomposition

$$\left(\rho^S\right)^\sharp = \sum_i \lambda_i |\Psi_i\rangle\langle\Psi_i| \ ,$$

where $\{\lambda_i\} \in \mathbb{R}$ are the eigenvalues of $\left(\rho^S\right)^\sharp \in \mathcal{H}^{SB}$ and each $|\Psi_i\rangle \in \mathcal{H}^{SB}$ can be written in terms of the system and bath basis states, i.e.,

$$|\Psi_i\rangle = \sum_{mn} a_{mn}^{(i)} |s_m b_n\rangle \ ,$$

with $\{a_{mn}\} \in \mathbb{C}$. The states $|s_m\rangle \in \mathcal{H}^S$ and $|b_n\rangle \in \mathcal{H}^B$. The initial composite state can then be written as

$$(\rho^S)^{\#} = \sum_i \lambda_i \left(\sum_{mn} a_{mn}^{(i)} |s_m b_n\rangle \right) \left(\sum_{m'n'} a_{m'n'}^{(i)*} \langle s_{m'} b_{n'}| \right) \tag{A.1}$$

$$= \sum_{imnm'n'} \lambda_i a_{mn}^{(i)} a_{m'n'}^{(i)*} |s_m b_n\rangle \langle s_{m'} b_{n'}| \tag{A.2}$$

$$\equiv \sum_{imnm'n'} \lambda_i a_{mn}^{(i)} a_{m'n'}^{(i)*} \left(|s_m\rangle \langle s_{m'}| \otimes |b_n\rangle \langle b_{n'}| \right) . \tag{A.3}$$

The composite evolution also has an eigendecomposition

$$U^{SB} = \sum_j v_j |\phi_j\rangle \langle \phi_j| ,$$

where $\{v_j\} \in \mathbb{C}$ are the eigenvalues of $U^{SB} \in \mathcal{B}(\mathcal{H}^{SB})$ and each $|\phi_i\rangle \in \mathcal{H}^{SB}$ can (again) be written in terms of the system and bath basis states, i.e.,

$$|\phi_j\rangle = \sum_{xy} c_{xy}^{(j)} |s_x b_y\rangle ,$$

with $\{c_{xy}\} \in \mathbb{C}$. The composite evolution is then written as

$$U^{SB} = \sum_{jxyx'y'} v_j c_{xy}^{(j)} c_{x'y'}^{(j)*} |s_x b_y\rangle \langle s_{x'} b_{y'}| \tag{A.4}$$

$$\equiv \sum_{jxyx'y'} v_j c_{xy}^{(j)} c_{x'y'}^{(j)*} \left(|s_x\rangle \langle s_{x'}| \otimes |b_y\rangle \langle b_{y'}| \right) . \tag{A.5}$$

Hence,

$$\left(U^{SB} \right)^{\dagger} = \sum_{kopo'p'} v_k c_{op}^{(k)*} c_{o'p'}^{(k)} \left(|s_{o'}\rangle \langle s_o| \otimes |b_{p'}\rangle \langle b_p| \right) ,$$

where $v_k = v_j^*$ in the sum of U^{SB}.

The reduced system dynamics are then written down as

$$\epsilon(\rho^S) = \left((U^{SB}) (\rho^S)^{\#} (U^{SB})^{\dagger} \right)^{\flat} ,$$

and plugging in everything from above yields

$$
\epsilon(\rho^S) = \left(\left(\sum_{jxyx'y'} \nu_j c_{xy}^{(j)} c_{x'y'}^{(j)*} \left(|s_x\rangle\langle s_{x'}| \otimes |b_y\rangle\langle b_{y'}|\right)\right)\right.
$$

$$
\left(\sum_{imnm'n'} \lambda_i a_{mn}^{(i)} a_{m'n'}^{(i)*} \left(|s_m\rangle\langle s_{m'}| \otimes |b_n\rangle\langle b_{n'}|\right)\right)
$$

$$
\left.\left(\sum_{kopo'p'} \nu_k c_{op}^{(k)*} c_{o'p'}^{(k)} \left(|s_{o'}\rangle\langle s_o| \otimes |b_{p'}\rangle\langle b_p|\right)\right)\right)^b
$$

$$
= \left(\sum_{\substack{jxyx' \\ imnm'n' \\ kopo'}} \lambda_i \nu_j \nu_k a_{mn}^{(i)} a_{m'n'}^{(i)*} c_{xy}^{(j)} c_{x'n}^{(j)*} c_{op}^{(k)*} c_{o'n'}^{(k)} \left(|s_x\rangle \langle s_{x'}|s_m\rangle \langle s_{m'}|s_{o'}\rangle \langle s_o| \otimes |b_y\rangle\langle b_p|\right)\right)^b
$$

$$
= \sum_{\substack{jxyx' \\ imnm'n' \\ kopo'}} \lambda_i \nu_j \nu_k a_{mn}^{(i)} a_{m'n'}^{(i)*} c_{xy}^{(j)} c_{x'n}^{(j)*} c_{op}^{(k)*} c_{o'n'}^{(k)} \operatorname{Tr}\left(|b_y\rangle\langle b_p|\right) \left(|s_x\rangle \langle s_{x'}|s_m\rangle \langle s_{m'}|s_{o'}\rangle \langle s_o|\right)
$$

$$
= \sum_{\substack{jxyx' \\ imnm'n' \\ kopo'}} \lambda_i \nu_j \nu_k a_{mn}^{(i)} a_{m'n'}^{(i)*} c_{xy}^{(j)} c_{x'n}^{(j)*} c_{op}^{(k)*} c_{o'n'}^{(k)}
$$

$$
\left(\sum_q \langle b_q|b_y\rangle\langle b_p|b_q\rangle\right) \left(|s_x\rangle \langle s_{x'}|s_m\rangle \langle s_{m'}|s_{o'}\rangle \langle s_o|\right)
$$

$$
= \sum_{\substack{jxx' \\ imnm'n' \\ qkoo'}} \lambda_i \nu_j \nu_k a_{mn}^{(i)} a_{m'n'}^{(i)*} c_{xq}^{(j)} c_{x'n}^{(j)*} c_{oq}^{(k)*} c_{o'n'}^{(k)} \left(|s_x\rangle \langle s_{x'}|s_m\rangle \langle s_{m'}|s_{o'}\rangle \langle s_o|\right)
$$

$$
= \sum_{imnm'n'q} \lambda_i a_{mn}^{(i)} a_{m'n'}^{(i)*} \hat{S}_{qn}|s_m\rangle\langle s_{m'}|\hat{S}_{qn'}^\dagger ,
$$

where $\langle b_\alpha|b_\beta\rangle = \delta_{\alpha\beta}$ is the delta function,

$$
\hat{S}_{qn} = \sum_{jxx'} \nu_j c_{xq}^{(j)} c_{x'n}^{(j)*} |s_x\rangle\langle s_{x'}|
$$

$$
= \left(I \otimes \langle b_q|\right) U^{SB} \left(I \otimes |b_n\rangle\right)
$$

and

$$
\hat{S}_{qn'}^\dagger = \sum_{koo'} \nu_k c_{oq}^{(k)*} c_{o'n'}^{(k)} |s_{o'}\rangle\langle s_o|
$$

$$
= \left(I \otimes \langle b_{n'}|\right) \left(U^{SB}\right)^\dagger \left(I \otimes |b_q\rangle\right) ,
$$

with I as the identity on the reduced system.

A.3   GENERAL FORM OF INITIAL REDUCED SYSTEM STATE

The initial state of the reduced system must be consistent, hence

$$\rho^S = \left(\left(\rho^S\right)^\sharp\right)^\flat ,$$

where $\rho^S \in \mathcal{S}(\mathcal{H}^S)$ and $(\rho^S)^\sharp \in \mathcal{S}(\mathcal{H}^{SB})$. The general form of the initial composite state $(\rho^S)^\sharp$ has been given in Appendix A.2 and can be plugged in to this consistency equation to yield

$$
\begin{aligned}
\rho^S &= \left(\sum_{imnm'n'} \lambda_i a_{mn}^{(i)} a_{m'n'}^{(i)*} \left(|s_m\rangle\langle s_{m'}| \otimes |b_n\rangle\langle b_{n'}| \right) \right)^\flat \\
&= \sum_{imnm'n'} \lambda_i a_{mn}^{(i)} a_{m'n'}^{(i)*} \, \mathrm{Tr}\left(|b_n\rangle\langle b_{n'}| \right) |s_m\rangle\langle s_{m'}| \\
&= \sum_{imnm'n'} \lambda_i a_{mn}^{(i)} a_{m'n'}^{(i)*} \delta_{nn'} |s_m\rangle\langle s_{m'}| \\
&= \sum_{imnm'} \lambda_i a_{mn}^{(i)} a_{m'n}^{(i)*} |s_m\rangle\langle s_{m'}| .
\end{aligned}
$$

If the initial state of the bath is some fixed pure state $|b_n\rangle\langle b_{n'}| = |b_\phi\rangle\langle b_\phi|$, then the initial state of the reduced system becomes

$$\rho^S = \sum_{imm'} \lambda_i a_{m\phi}^{(i)} a_{m'\phi}^{(i)*} |s_m\rangle\langle s_{m'}| .$$

A.4   COMPLETENESS RELATION FOR $\hat{S}_{q\phi}$

From Appendix A.2, the operators are given as

$$
\begin{aligned}
\hat{S}_{qn} &= \sum_{jxx'} v_j c_{xq}^{(j)} c_{x'n}^{(j)*} |s_x\rangle\langle s_{x'}| \\
&= \left(I \otimes \langle b_q| \right) U^{SB} \left(I \otimes |b_n\rangle \right)
\end{aligned}
$$

and

$$
\begin{aligned}
\hat{S}_{qn'}^\dagger &= \sum_{koo'} v_k c_{oq}^{(k)*} c_{o'n'}^{(k)} |s_{o'}\rangle\langle s_o| \\
&= \left(I \otimes \langle b_{n'}| \right) \left(U^{SB} \right)^\dagger \left(I \otimes |b_q\rangle \right) .
\end{aligned}
$$

From these definitions,

$$
\begin{aligned}
\sum_q \hat{S}_{qn'}^{\dagger}\hat{S}_{qn} &= \sum_q (I \otimes \langle b_{n'}|)\left(U^{SB}\right)^{\dagger}\left(I \otimes |b_q\rangle\right)\left(I \otimes \langle b_q|\right)U^{SB}\left(I \otimes |b_n\rangle\right) \\
&= \sum_q (I \otimes \langle b_{n'}|)\left(U^{SB}\right)^{\dagger}\left(I \otimes |b_q\rangle\langle b_q|\right)U^{SB}\left(I \otimes |b_n\rangle\right) \\
&= (I \otimes \langle b_{n'}|)\left(U^{SB}\right)^{\dagger}\left(I \otimes \left(\sum_q |b_q\rangle\langle b_q|\right)\right)U^{SB}\left(I \otimes |b_n\rangle\right) \\
&= (I \otimes \langle b_{n'}|)\left(U^{SB}\right)^{\dagger}\left(I \otimes I\right)U^{SB}\left(I \otimes |b_n\rangle\right) \\
&= (I \otimes \langle b_{n'}|)\left(U^{SB}\right)^{\dagger}U^{SB}\left(I \otimes |b_n\rangle\right) \\
&= (I \otimes \langle b_{n'}|)\left(I \otimes |b_n\rangle\right) \\
&= I\delta_{n'n} ,
\end{aligned}
$$

where $\left(U^{SB}\right)^{\dagger}U^{SB} = I$ by definition and $\langle b_{n'}|b_n\rangle = \delta_{n'n}$ is the delta function.

This result can also be seen as follows:

$$
\begin{aligned}
\sum_q \hat{S}_{qn'}^{\dagger}\hat{S}_{qn} &= \sum_q \left(\sum_{koo'} v_k c_{oq}^{(k)*}c_{o'n'}^{(k)}|s_{o'}\rangle\langle s_o|\right)\left(\sum_{jxx'} v_j c_{xq}^{(j)}c_{x'n}^{(j)*}|s_x\rangle\langle s_{x'}|\right) \\
&= \sum_{\substack{qkoo' \\ jxx'}} v_k c_{oq}^{(k)*}c_{o'n'}^{(k)}v_j c_{xq}^{(j)}c_{x'n}^{(j)*}\delta_{ox}|s_{o'}\rangle\langle s_{x'}| \\
&= \sum_{\substack{qko' \\ jxx'}} v_k c_{xq}^{(k)*}c_{o'n'}^{(k)}v_j c_{xq}^{(j)}c_{x'n}^{(j)*}|s_{o'}\rangle\langle s_{x'}| \\
&= \sum_{\substack{ko' \\ jx'}} v_k c_{o'n'}^{(k)}v_j c_{x'n}^{(j)*}\left(\sum_{qx} c_{xq}^{(k)*}c_{xq}^{(j)}\right)|s_{o'}\rangle\langle s_{x'}| .
\end{aligned}
$$

Notice, from Appendix A.2,

$$
\begin{aligned}
\left(U^{SB}\right)^{\dagger}U^{SB} &= \left(\sum_{kopo'p'} v_k c_{op}^{(k)*}c_{o'p'}^{(k)}\left(|s_{o'}\rangle\langle s_o| \otimes |b_{p'}\rangle\langle b_p|\right)\right) \\
&\quad \left(\sum_{jxyx'y'} v_j c_{xy}^{(j)}c_{x'y'}^{(j)*}\left(|s_x\rangle\langle s_{x'}| \otimes |b_y\rangle\langle b_{y'}|\right)\right) \\
&= \sum_{\substack{kopo'p' \\ jxyx'y'}} v_k c_{op}^{(k)*}c_{o'p'}^{(k)}v_j c_{xy}^{(j)}c_{x'y'}^{(j)*}\delta_{ox}\delta_{py}\left(|s_{o'}\rangle\langle s_{x'}| \otimes |b_{p'}\rangle\langle b_{y'}|\right) \\
&= \sum_{\substack{ko'p' \\ jxyx'y'}} v_k c_{xy}^{(k)*}c_{o'p'}^{(k)}v_j c_{xy}^{(j)}c_{x'y'}^{(j)*}\left(|s_{o'}\rangle\langle s_{x'}| \otimes |b_{p'}\rangle\langle b_{y'}|\right)
\end{aligned}
$$

and

$$
\begin{aligned}
\left(\left(U^{SB}\right)^\dagger U^{SB}\right)^\flat &= \left(\sum_{\substack{ko'p' \\ jxyx'y'}} v_k c_{xy}^{(k)*} c_{o'p'}^{(k)} v_j c_{xy}^{(j)} c_{x'y'}^{(j)*} \left(|s_{o'}\rangle\langle s_{x'}| \otimes |b_{p'}\rangle\langle b_{y'}| \right) \right)^\flat \\
&= \sum_{\substack{ko'p' \\ jxyx'y'}} v_k c_{xy}^{(k)*} c_{o'p'}^{(k)} v_j c_{xy}^{(j)} c_{x'y'}^{(j)*} \delta_{p'y'} |s_{o'}\rangle\langle s_{x'}| \\
&= \sum_{\substack{ko'jx \\ yx'p'}} v_k c_{xy}^{(k)*} c_{o'p'}^{(k)} v_j c_{xy}^{(j)} c_{x'p'}^{(j)*} |s_{o'}\rangle\langle s_{x'}| \\
&= \sum_{\substack{ko'j \\ x'p'}} v_k c_{o'p'}^{(k)} v_j c_{x'p'}^{(j)*} \left(\sum_{xy} c_{xy}^{(k)*} c_{xy}^{(j)} \right) |s_{o'}\rangle\langle s_{x'}| \ .
\end{aligned}
$$

Hence,

$$
\left(\left(U^{SB}\right)^\dagger U^{SB}\right)^\flat = \sum_{p'} \left(\sum_q \hat{S}_{qn'}^\dagger \hat{S}_{qn} \right) \delta_{n'p'}\delta_{np'} = \sum_n \left(\sum_q \hat{S}_{qn'}^\dagger \hat{S}_{qn} \right) \delta_{n'n} \ .
$$

Notice

$$
\left(U^{SB}\right)^\dagger U^{SB} = I \in \mathcal{H}^{SB} \Rightarrow \left(\left(U^{SB}\right)^\dagger U^{SB}\right)^\flat = I \in \mathcal{B}(\mathcal{H}^S) \ ,
$$

where I is the identity operator, which implies

$$
\sum_{nq} \hat{S}_{qn'}^\dagger \hat{S}_{qn} \delta_{n'n} = I \ .
$$

This relation is expected from the definition of the operators $\hat{S}_{qn}$ as the re-stacked Choi representation of the channel, which is a Hermitian matrix [27].

If the bath is in some specific pure state $|b_n\rangle\langle b_{n'}| = |b_\phi\rangle\langle b_\phi|$, then there is no sum over the bath states. In such a case,

$$
\sum_q \hat{S}_{q\phi}^\dagger \hat{S}_{q\phi} = I \ .
$$

A.5   RABI MODEL DERIVATION

Maxwell's equations relate the electric and magnetic fields, $\mathbf{E}$ and $\mathbf{B}$, to the charge density ϱ and the current density $\mathbf{j}$ [29]. They are

$$\nabla \cdot \mathbf{E}(\mathbf{r}, t) = \frac{\varrho(\mathbf{r}, t)}{\epsilon_0}$$
$$\nabla \cdot \mathbf{B}(\mathbf{r}, t) = 0$$
$$\nabla \times \mathbf{E}(\mathbf{r}, t) = -\frac{\partial}{\partial t}\mathbf{B}(\mathbf{r}, t)$$
$$\nabla \times \mathbf{B}(\mathbf{r}, t) = \frac{1}{c^2}\frac{\partial}{\partial t}\mathbf{E}(\mathbf{r}, t) + \frac{\mathbf{j}(\mathbf{r}, t)}{\epsilon_0 c^2} \; ,$$

where the constant c is the speed of light in a vacuum and the constant ϵ_0 is the vacuum permittivity. The dependence on the position vector $\mathbf{r}$ and time t has been written explicitly for the electric and magnetic field vectors, the current density vector, and the charge density. The middle two equations from above suggest the following forms for the electric and magnetic fields:

$$\mathbf{B}(\mathbf{r}, t) = \nabla \times \mathbf{A}(\mathbf{r}, t)$$
$$\mathbf{E}(\mathbf{r}, t) = -\frac{\partial}{\partial t}\mathbf{A}(\mathbf{r}, t) - \nabla \Phi(\mathbf{r}, t) \; ,$$

where $\mathbf{A}$ is called the "vector potential field" and Φ is called the "scalar potential field."

The Lorentz force law is

$$\mathbf{F} = q\mathbf{E} + q\,(\mathbf{v} \times \mathbf{B})$$

where q is the electric charge of a particle moving with velocity $\mathbf{v}$ through the electric and magnetic fields $\mathbf{E}$ and $\mathbf{B}$. The time and space dependence of the fields is left out of these expressions for clarity of presentation, but it should be remembered that those dependences are still implied. This expression can be written in terms of the potentials as

$$\mathbf{F} = q\left(-\nabla \Phi + \nabla\,(\mathbf{v} \cdot \mathbf{A}) - \frac{d}{dt}\mathbf{A}\right)$$

where the vector identity (for some vectors $\mathbf{K_1}$ and $\mathbf{K_2}$)

$$\mathbf{K_1} \times (\nabla \times \mathbf{K_2}) = \nabla\,(\mathbf{K_1} \cdot \mathbf{K_2}) - (\mathbf{K_1} \cdot \nabla)\,\mathbf{K_2} - (\mathbf{K_2} \cdot \nabla)\,\mathbf{K_1} - \mathbf{K_2} \times (\nabla \times \mathbf{K_1})$$

is used along with the fact that $\nabla \mathbf{v} = 0$ and $\nabla \times \mathbf{v} = \mathbf{0}$ to yield

$$\mathbf{v} \times (\nabla \times \mathbf{A}) = \nabla\,(\mathbf{v} \cdot \mathbf{A}) - (\mathbf{v} \cdot \nabla)\,\mathbf{A} \; ,$$

which can be rewritten by recognizing

$$\frac{d}{dt}\mathbf{A} = \frac{\partial}{\partial t}\mathbf{A} + \frac{\partial}{\partial x}\mathbf{A}\frac{d}{dt}x + \frac{\partial}{\partial y}\mathbf{A}\frac{d}{dt}y + \frac{\partial}{\partial z}\mathbf{A}\frac{d}{dt}z = \frac{\partial}{\partial t}\mathbf{A} + (\mathbf{v} \cdot \nabla)\,\mathbf{A} \; .$$

The system under investigation will be assumed to be conservative (i.e., $\mathbf{F} = -\nabla V$ where V is the potential energy), which allows us the write the potential energy as

$$V = q\Phi - q(\mathbf{v} \cdot \mathbf{A}) \ .$$

The kinetic energy of the system is given by

$$T = \frac{m\mathbf{v} \cdot \mathbf{v}}{2} \ ,$$

which implies the (classical) electromagnetic Lagrangian is

$$L = T - V = \frac{m\mathbf{v} \cdot \mathbf{v}}{2} - q\Phi + q(\mathbf{v} \cdot \mathbf{A}) \ .$$

The canonical momentum

$$\mathbf{p} = \frac{\partial}{\partial \mathbf{v}} L = m\mathbf{v} + q\mathbf{A}$$

leads to the (classical) electromagnetic Hamiltonian

$$H = \mathbf{p} \cdot \mathbf{v} - L = \frac{1}{2m}(\mathbf{p} - q\mathbf{A}(\mathbf{r},t))^2 + q\Phi(\mathbf{r},t) \ ,$$

where the explicit dependence of the potentials on position and time has again been added for emphasis. Notice that this Hamiltonian is for a single particle with a potential governed solely by the electromagnetic field (i.e., there are no other potentials in this system).

In standard quantum mechanics, the correspondence principle is used to form a quantum Hamiltonian from the above classical one. The operators that satisfy the canonical commutation relations are $\hat{\mathbf{r}}$ and $\hat{\mathbf{p}} = -i\hbar\nabla$ (see [29] for a discussion of these points with respect to the electromagnetic Hamiltonian). Using this substitution, the quantum Hamiltonian can immediately be written down as

$$H_e = \frac{1}{2m}\left(-i\hbar\nabla - e\hat{\mathbf{A}}(\hat{\mathbf{r}},t)\right)^2 - e\Phi(\hat{\mathbf{r}},t) \ ,$$

where we have assumed that the charged particle is an electron, i.e., $q = e$ where e as the charge of the electron.

The classical fields are said to invariant under "gauge transformations" of the form

$$\mathbf{A}(\mathbf{r},t) \rightarrow \mathbf{A}'(\mathbf{r},t) = \mathbf{A}(\mathbf{r},t) + \nabla F(\mathbf{r},t)$$
$$\Phi(\mathbf{r},t) \rightarrow \Phi'(\mathbf{r},t) = \Phi(\mathbf{r},t) - \frac{\partial}{\partial t} F(\mathbf{r},t) \ ,$$

where $F(\mathbf{r},t)$ is an arbitrary function of the position and time. The idea is that the same fields $\mathbf{E}$ and $\mathbf{B}$ can be described by either $\mathbf{A}(\mathbf{r},t)$ and $\Phi(\mathbf{r},t)$ or $\mathbf{A}'(\mathbf{r},t)$ and $\Phi'(\mathbf{r},t)$. This freedom can be useful in working through the above equations. The "Coulomb" (or radiation [29]) gauge is the condition

$$\nabla \cdot \mathbf{A}(\mathbf{r},t) = 0 \ .$$

Applying this gauge to the quantum electromagnetic Hamiltonian yields

$$H_e = -\frac{\hbar^2}{2m}\nabla^2 + \frac{i\hbar e}{m}\hat{\mathbf{A}}\left(\hat{\mathbf{r}}, t\right) \cdot \nabla + \frac{e^2}{2m}\hat{\mathbf{A}}^2\left(\hat{\mathbf{r}}, t\right) - e\Phi(\hat{\mathbf{r}}, t) \ .$$

The scalar potential can be set to zero (sometimes referred to as the "velocity gauge" [33]), and we can assume the fields are "weak" in the sense that the terms proportional to the square of the vector potential can be ignored.[1] These assumptions lead to

$$H_e \approx -\frac{\hbar^2}{2m}\nabla^2 + \frac{i\hbar e}{m}\hat{\mathbf{A}}\left(\hat{\mathbf{r}}, t\right) \cdot \nabla \ .$$

Ehrenfest's theorem [84] states

$$\frac{d}{dt}\langle O \rangle = \frac{i}{\hbar}\langle [H, O] \rangle + \langle \frac{\partial}{\partial t} O \rangle$$

for some operator O. In particular, for the position operator,

$$\frac{d}{dt}\langle \hat{\mathbf{r}} \rangle = \frac{i}{\hbar}\langle [H, \hat{\mathbf{r}}] \rangle + \langle \frac{\partial}{\partial t}\hat{\mathbf{r}} \rangle \ .$$

The position operator does not explicitly depend on time, so this equation reduces to

$$\frac{d}{dt}\langle \hat{\mathbf{r}} \rangle = \frac{i}{\hbar}\langle [\frac{\hat{\mathbf{p}}^2}{2m} + V(\hat{\mathbf{r}}), \hat{\mathbf{r}}] \rangle$$

where the Hamiltonian operator has been written out in terms of the potential and momentum as $H = \frac{\hat{\mathbf{p}}^2}{2m} + V(\hat{\mathbf{r}})$. Notice $\langle [V(\hat{\mathbf{r}}), \hat{\mathbf{r}}] \rangle = 0$, so everything can be further reduced:

$$\begin{aligned}
\frac{d}{dt}\langle \hat{\mathbf{r}} \rangle &= \frac{i}{\hbar}\langle [\frac{\hat{\mathbf{p}}^2}{2m}, \hat{\mathbf{r}}] \rangle \\
&= \frac{i}{2m\hbar}\langle [\hat{\mathbf{p}}^2, \hat{\mathbf{r}}] \rangle \\
&= \frac{i}{2m\hbar}\langle \hat{\mathbf{p}}[\hat{\mathbf{p}}, \hat{\mathbf{r}}] + [\hat{\mathbf{p}}, \hat{\mathbf{r}}]\hat{\mathbf{p}} \rangle \\
&= -\frac{i}{2m\hbar}\langle \hat{\mathbf{p}} i\hbar + i\hbar\hat{\mathbf{p}} \rangle \\
&= \frac{2\hbar}{2m\hbar}\langle \hat{\mathbf{p}} \rangle \\
&= \frac{\langle \hat{\mathbf{p}} \rangle}{m} \ .
\end{aligned}$$

This result implies

$$\langle \hat{\mathbf{p}} \rangle = m\frac{d\langle \hat{\mathbf{r}} \rangle}{dt} \ ,$$

---

[1]See pp. 688–689 in [62] for a discussion of why the assumption of "weak" fields is reasonable "for all but the most intense optical fields."

and this final result will be used as a justification for making the substitution $-i\hbar\nabla \to m\frac{d\hat{\mathbf{r}}}{dt}$, which is referred to as the "time-dependent form" of the momentum operator. Making this substitution yields

$$H_e \approx -\frac{\hbar^2}{2m}\nabla^2 + e\hat{\mathbf{A}}\left(\hat{\mathbf{r}},t\right) \cdot \frac{d}{dt}\hat{\mathbf{r}} \ .$$

Notice

$$\hat{\mathbf{A}}(\hat{\mathbf{r}},t) \cdot \frac{d\hat{\mathbf{r}}}{dt} = \frac{d}{dt}\left(\hat{\mathbf{A}}(\hat{\mathbf{r}},t) \cdot \hat{\mathbf{r}}\right) - \frac{d\hat{\mathbf{A}}(\hat{\mathbf{r}},t)}{dt} \cdot \hat{\mathbf{r}} \ ,$$

which leads to

$$H_e \approx -\frac{\hbar^2}{2m}\nabla^2 + e\left(\frac{d}{dt}\left(\hat{\mathbf{A}}(\hat{\mathbf{r}},t) \cdot \hat{\mathbf{r}}\right) - \frac{d\hat{\mathbf{A}}(\hat{\mathbf{r}},t)}{dt} \cdot \hat{\mathbf{r}}\right) \ .$$

At this point it will be further assumed that the electromagnetic field has no spatial variation across the qubit. This is the assumption that the position operator $\hat{\mathbf{r}}$ in the potentials can be treated as a constant and replaced by its average value $\mathbf{r}_0$. Specifically, $\hat{\mathbf{A}}\left(\hat{\mathbf{r}},t\right) \to \mathbf{A}\left(\mathbf{r}_0,t\right)$. This assumption is referred to as the "dipole approximation."

Consider a specific form of the classical vector potential given as

$$\mathbf{A}(\mathbf{r},t) = \frac{1}{2}\left(\mathbf{B}(t) \times \mathbf{r}\right) \ ,$$

where the magnetic field does not explicitly depend on position. The vector identity (for some vectors $\mathbf{K_1}$ and $\mathbf{K_2}$)

$$\nabla \times (\mathbf{K_1} \times \mathbf{K_2}) = \mathbf{K_1}\left(\nabla \cdot \mathbf{K_2}\right) - \mathbf{K_2}\left(\nabla \cdot \mathbf{K_1}\right) + (\mathbf{K_2} \cdot \nabla)\mathbf{K_1} - (\mathbf{K_1} \cdot \nabla)\mathbf{K_2} \ ,$$

along with the fact that $\nabla \cdot \mathbf{r} = 3$ and $\nabla\mathbf{r} = 1$, implies

$$\nabla \times \mathbf{A}(\mathbf{r},t) = \mathbf{B}(t)$$

as desired. The vector identities

$$\nabla \cdot (\mathbf{K_1} \times \mathbf{K_2}) = \mathbf{K_2} \cdot (\nabla \times \mathbf{K_1}) - \mathbf{K_1} \cdot (\nabla \times \mathbf{K_2})$$

and

$$\nabla \times (\nabla \times \mathbf{K_1}) = \nabla(\nabla \cdot \mathbf{K_1}) - \nabla^2\mathbf{K_1}$$

imply

$$\nabla \cdot \mathbf{A}(\mathbf{r},t) = 0$$

because $\nabla^2\mathbf{A}(\mathbf{r},t) = 0$ in the dipole approximation and we are working in the Coulomb gauge (which is consistent with the above result). It follows that this specific form of the vector potential is valid in the Coulomb gauge given the dipole approximation.

The first vector potential term in the Hamiltonian can be shown to be zero after applying the dipole approximation and the above form of the vector potential, i.e.,

$$\frac{d}{dt}\left(\mathbf{A}(\mathbf{r}_0, t) \cdot \hat{\mathbf{r}}\right) = 0$$

where we have used the $\mathbf{K}_1 \cdot \mathbf{K}_2 = \mathbf{K}_2 \cdot \mathbf{K}_1$ property of the scalar product, the triple scalar product rule (i.e., $\mathbf{K}_1 \cdot (\mathbf{K}_2 \times \mathbf{K}_3) = \mathbf{K}_2 \cdot (\mathbf{K}_3 \times \mathbf{K}_1) = \mathbf{K}_3 \cdot (\mathbf{K}_1 \times \mathbf{K}_2)$ for some vectors $\mathbf{K}_1, \mathbf{K}_2$, and $\mathbf{K}_3$), and the fact that $\mathbf{r} \times \mathbf{r} = 0$.

The other vector potential term in the Hamiltonian can also be reduced using the dipole approximation and remembering that the scalar potential has already been assumed to be zero. As such, the electric field is given only by the negative time derivative of the vector potential and

$$-\frac{d\,\mathbf{A}(\mathbf{r}_0, t)}{dt} \cdot \hat{\mathbf{r}} = e\mathbf{E}\,(\mathbf{r}_0, t) \cdot \hat{\mathbf{r}} \ .$$

The final approximate Hamiltonian is

$$H_e = -\frac{\hbar^2}{2m}\nabla^2 + e\mathbf{E}(\mathbf{r}_0, t) \cdot \hat{\mathbf{r}} = H_f + H_{di} \ ,$$

where the Hamiltonian is explicitly split into two terms in the last step with

$$H_{di} \equiv e\mathbf{E}(\mathbf{r}_0, t) \cdot \hat{\mathbf{r}}$$

and

$$H_f \equiv -\frac{\hbar^2}{2m}\nabla^2$$

to emphasize that the Hamiltonian is just the free particle Hamiltonian with a "dipole term" added.

The dipole Hamiltonian can be written as

$$H_{di} = -\mathbf{E}(\mathbf{r}_0, t) \cdot \hat{\mathbf{D}}$$

where the dipole operator is defined as $\hat{\mathbf{D}} = -e\hat{\mathbf{r}}$. Although this notation is not used much in our work, it is the standard notation of the dipole interaction Hamiltonian in most derivations [29, 60, 62].

We will now consider a two level system with energy levels labeled $|0\rangle$ and $|1\rangle$ that have associated energy eigenvalues of $-\hbar\omega_0/2$ and $\hbar\omega_0/2$ respectively. The system interacts with an electromagnetic field that is propagating in the direction $\mathbf{n}$, oscillates with a frequency ω, and has a polarization of ϵ, which can be described as

$$\mathbf{E}(\mathbf{r}_0, t) = E_0\left(\epsilon e^{i(\omega t - (\mathbf{k} \cdot \mathbf{n})|\mathbf{r}_0|)} + \epsilon^* e^{-i(\omega t - (\mathbf{k} \cdot \mathbf{n})|\mathbf{r}_0|)}\right) \ ,$$

where $E_0 \in \mathbb{R}$ is the amplitude of the field. It can be assumed that the two level system is located at the origin. Therefore, $|\mathbf{r}_0| = 0$, and the second terms in the exponentials vanish leaving only

$$\mathbf{E}(t) = E_0 \left(\epsilon e^{i\omega t} + \epsilon^* e^{-i\omega t} \right) .$$

This field can be used to write down the final approximate Hamiltonian in matrix form.

The first step is to find the off-diagonal terms of the dipole Hamiltonian H_{di} as

$$\langle 1 | H_{di} | 0 \rangle = \langle 1 | (e\mathbf{E}(t) \cdot \hat{\mathbf{r}}) | 0 \rangle = e E_0 \hat{\mathbf{r}}_{10} \cdot \left(\epsilon e^{i\omega t} + \epsilon^* e^{-i\omega t} \right) ,$$

with $\hat{\mathbf{r}}_{10} = \langle 1 | \hat{\mathbf{r}} | 0 \rangle$. The Hamiltonian is Hermitian by definition, hence $\langle 0 | H_{di} | 1 \rangle = (\langle 1 | H_{di} | 0 \rangle)^*$. In general, $\langle 0 | H_{di} | 1 \rangle$ is a complex quantity [33], but it will be real for transitions between bound states [60] (where the states $|0\rangle$ and $|1\rangle$ are real), which is precisely the situation we wish to model in this derivation. As such, we will use $\langle 0 | H_{di} | 1 \rangle = \langle 1 | H_{di} | 0 \rangle$ and ignore the customary explicit conjugation notation in the Hamiltonians that follow.

The position operator $\hat{\mathbf{r}}$ has odd parity. To see this fact, define the parity operator $\hat{\mathbf{P}}$ by its action on the position state of a system: $\hat{\mathbf{P}} | r \rangle = | -r \rangle$ and $\langle r | \hat{\mathbf{P}}^\dagger = \langle -r |$. Notice $\hat{\mathbf{P}}^2 = \hat{\mathbf{P}}^\dagger \hat{\mathbf{P}} = \hat{\mathbf{P}} \hat{\mathbf{P}}^\dagger = I$ (where I is the identity operator) and $\hat{\mathbf{P}} = \hat{\mathbf{P}}^\dagger$. The position operator $\hat{\mathbf{r}}$ is defined by $\hat{\mathbf{r}} | r \rangle = r' | r \rangle$ which implies

$$\begin{aligned} \hat{\mathbf{P}} \hat{\mathbf{r}} \hat{\mathbf{P}} | r \rangle &= \hat{\mathbf{P}} \hat{\mathbf{r}} | -r \rangle \\ &= \hat{\mathbf{P}} (-r') | -r \rangle \\ &= (-r') | r \rangle \\ &= -(\hat{\mathbf{r}} | r \rangle) . \end{aligned}$$

Therefore, $\hat{\mathbf{r}}$ has odd parity. This result implies

$$\begin{aligned} \langle 0 | \hat{\mathbf{r}} | 0 \rangle &= \langle 0 | \hat{\mathbf{P}}^\dagger \hat{\mathbf{P}} \hat{\mathbf{r}} \hat{\mathbf{P}}^\dagger \hat{\mathbf{P}} | 0 \rangle \\ &= - \langle 0 | \hat{\mathbf{P}}^\dagger \hat{\mathbf{r}} \hat{\mathbf{P}} | 0 \rangle \\ &= - \langle 0 | \hat{\mathbf{r}} | 0 \rangle \end{aligned}$$

because the parity of the state $|0\rangle$ is either even or odd (i.e., the state $|0\rangle$ is an eigenstate of the parity operator with an eigenvalue of either $+1$ or -1). The final implication is that $\langle 0 | \hat{\mathbf{r}} | 0 \rangle = 0$. Similarly, $\langle 1 | \hat{\mathbf{r}} | 1 \rangle = 0$. This result, in turn, implies $\langle 0 | H_{di} | 0 \rangle = \langle 1 | H_{di} | 1 \rangle = 0$; hence, there are no diagonal terms for H_{di}.

Putting everything together yields the matrix form of the final approximate Hamiltonian as [51]

$$H_e = \begin{pmatrix} -\dfrac{\hbar\omega_o}{2} & e E_0 \hat{\mathbf{r}}_{10} \cdot \left(\epsilon e^{i\omega t} + \epsilon^* e^{-i\omega t} \right) \\ e E_0 \hat{\mathbf{r}}_{10} \cdot \left(\epsilon e^{i\omega t} + \epsilon^* e^{-i\omega t} \right) & \dfrac{\hbar\omega_o}{2} \end{pmatrix} .$$

Several approximations have already been applied to derive this Hamiltonian, but it is still too complicated to solve Schrödinger's equation directly because of the time dependence in the off-diagonal terms. As a final step, we apply the rotating wave approximation to yield a Hamiltonian more amenable to theoretical investigation.

The first step is to transform the wavefunction to a frame that rotates at the frequency of the optical field (i.e., ω) using

$$U = \mathrm{diag}\left(e^{i\omega t/2}, e^{-i\omega t/2}\right) \ .$$

The wavefunction in this new frame is defined as $|\psi'\rangle = U|\psi\rangle$ (or $|\psi\rangle = U^\dagger|\psi'\rangle$) and Schrödinger's equation becomes

$$H_e|\psi\rangle = -i\hbar\frac{d}{dt}|\psi\rangle \rightarrow H_e U^\dagger|\psi'\rangle = -i\hbar\frac{d}{dt}\left(U^\dagger|\psi'\rangle\right)$$
$$= -i\hbar U^\dagger\frac{d}{dt}|\psi'\rangle - i\hbar\left(\frac{d}{dt}U^\dagger\right)|\psi'\rangle \ ,$$

which can be multiplied by U (and rearranged) to yield

$$\left(UH_e U^\dagger + i\hbar U\frac{d}{dt}U^\dagger\right)|\psi'\rangle = -i\hbar\frac{d}{dt}|\psi'\rangle$$
$$H_e'|\psi'\rangle = -i\hbar\frac{d}{dt}|\psi'\rangle$$

where the new "rotated" Hamiltonian can now be written down directly as

$$H_e' = \left(UH_e U^\dagger + i\hbar U\frac{d}{dt}U^\dagger\right)$$
$$= \begin{pmatrix} -\frac{\hbar(\omega_o - \omega)}{2} & eE_0\mathbf{r}_{10}\cdot\left(\epsilon + \epsilon^* e^{2i\omega t}\right) \\ eE_0\mathbf{r}_{10}\cdot\left(\epsilon + \epsilon^* e^{-2i\omega t}\right) & \frac{\hbar(\omega_o - \omega)}{2} \end{pmatrix} \ .$$

The final assumption in this derivation is to assume that the external field is close to resonance with the energy gap of the atom, i.e., $(\omega - \omega_0) << \omega$ and $E_0 << \omega$, which means the fast oscillating term of $e^{2i\omega t}$ can be ignored in the Hamiltonian. All of this work leads to

$$H_e' = \begin{pmatrix} -\frac{\hbar(\omega_o - \omega)}{2} & eE_0\left(\mathbf{r}_{10}\cdot\epsilon\right) \\ eE_0\left(\mathbf{r}_{10}\cdot\epsilon\right) & \frac{\hbar(\omega_o - \omega)}{2} \end{pmatrix}$$
$$= \frac{\hbar}{2}\begin{pmatrix} -\nu & \Omega \\ \Omega & \nu \end{pmatrix} \ ,$$

where

$$\nu \equiv \omega_o - \omega$$

is called the detuning and

$$\Omega \equiv \frac{2eE_0\left(\mathbf{r}_{10}\cdot\epsilon\right)}{\hbar}$$

is called the Rabi frequency. Notice that the Rabi frequency is always real in our derivation. This model of an atom is called the Rabi model, and it is a common model used in theoretical descriptions of nuclear magnetic resonance (NMR) and other quantum information experiments [67].

Bibliography

[1] D. Aharonov and M. Ben-Or. Fault-tolerant quantum computation with constant error rate. *SIAM Journal on Computing*, 38(4):1207–1282, 2008. DOI: 10.1137/S0097539799359385. 127

[2] D. Aharonov, A. Kitaev, and N. Nisan. Quantum circuits with mixed states. In *Proceedings of the thirtieth annual ACM symposium on Theory of computing*, STOC '98, pages 20–30, New York, NY, USA, 1998. ACM. DOI: 10.1145/276698.276708. 40, 121

[3] R. Alicki. Comment on "reduced dynamics need not be completely positive". *Phys. Rev. Lett.*, 75:3020–3020, Oct 1995. DOI: 10.1103/PhysRevLett.75.3020. xiii, 10, 13

[4] R. Alicki and M. Fannes. *Quantum Dynamical Systems*. Oxford University Press, 2001. DOI: 10.1093/acprof:oso/9780198504009.001.0001. xiii, 12, 139

[5] R. Alicki and K. Lendi. Recent developments. In *Quantum Dynamical Semigroups and Applications*, volume 717 of *Lecture Notes in Physics*, pages 109–121. Springer Berlin Heidelberg, 2007. DOI: 10.1007/3-540-70861-8_3. 139

[6] S. Barnett and P. Radmore. *Methods in Theoretical Quantum Optics*. Oxford Series in Optical and Imaging Sciences. Oxford University Press, Incorporated, 2002. DOI: 10.1093/acprof:oso/9780198563617.001.0001. 9, 95, 96

[7] A. S. Barzegar. *Open Quantum Systems and Error Correction*. PhD thesis, University of Southern California, 2009. 127, 140

[8] A. Ben-Aroya and A. Ta-Shma. On the complexity of approximating the diamond norm. *arXiv pre-print (quant-ph)*, arXiv:0902.3397, 2009. 121

[9] F. Benatti and R. Floreanini. Complete positivity and the k-k system. *Physics Letters B*, 389(1):100–106, 1996. DOI: 10.1016/S0370-2693(96)01228-2. 143

[10] F. Benatti and R. Floreanini. Completely positive dynamical maps and the neutral kaon system. *Nuclear Physics B*, 488(12):335–363, 1997. DOI: 10.1016/S0550-3213(96)00712-2. 143

[11] F. Benatti and R. Floreanini. Testing complete positivity. *Modern Physics Letters A*, 12(20):1465–1472, 1997. DOI: 10.1142/S0217732397001497. 143

[12] F. Benatti and R. Floreanini. Completely positive dynamics of correlated neutral kaons. *Nuclear Physics B*, 511(3):550–576, 1998. DOI: 10.1016/S0550-3213(97)00705-0. 143

[13] F. Benatti and R. Floreanini. Complete positivity and neutron interferometry. *Physics Letters B*, 451(34):422–429, 1999. DOI: 10.1016/S0370-2693(99)00177-X. 143

[14] F. Benatti and R. Floreanini. Open quantum dynamics: Complete positivity and entanglement. *International Journal of Modern Physics B*, 19(19):3063–3139, 2005. DOI: 10.1142/S0217979205032097. xiii, 45

[15] F. Bloch. Nuclear induction. *Phys. Rev.*, 70:460–474, Oct 1946. DOI: 10.1103/PhysRev.70.460. 39, 126

[16] K. Blum. *Density Matrix Theory and Applications*. Physics of Atoms and Molecules. Springer, 1996. DOI: 10.1007/978-1-4757-4931-1. 4

[17] N. Boulant, J. Emerson, T. F. Havel, D. G. Cory, and S. Furuta. Incoherent noise and quantum information processing. *The Journal of Chemical Physics*, 121(7):2955–2961, 2004. DOI: 10.1063/1.1773161. xiv, 49, 125

[18] H. Breuer and F. Petruccione. *The Theory of Open Quantum Systems*. Oxford University Press, 2007. DOI: 10.1093/acprof:oso/9780199213900.001.0001. 3, 6, 46, 110, 112, 143

[19] A. Brodutch, A. Datta, K. Modi, A. Rivas, and C. A. Rodriguez-Rosario. Vanishing quantum discord is not necessary for completely positive maps. *Phys. Rev. A*, 87:042301, Apr 2013. DOI: 10.1103/PhysRevA.87.042301. 58

[20] J. Budimir and J. Skinner. On the relationship between t1 and t2 for stochastic relaxation models. *Journal of Statistical Physics*, 49:1029–1042, 1987. DOI: 10.1007/BF01017558. 127

[21] P. Busch and P. Lahti. Completely positive mappings in quantum dynamics and measurement theory. *Foundations of Physics*, 20(12):1429–1439, 1990. DOI: 10.1007/BF01883516. 143

[22] F. Byron and R. Fuller. *Mathematics of Classical and Quantum Physics*. Dover Books on Physics Series. Dover Publ., 1992. 5

[23] E. Carlen. Trace inequalities and quantum entropy: an introductory course. In D. Ueltschi, editor, *Entropy and the Quantum*, Contemporary mathematics. American Mathematical Society, 2010. 9, 105

[24] H. A. Carteret, D. R. Terno, and K. Zyczkowski. Dynamics beyond completely positive maps: Some properties and applications. *Phys. Rev. A*, 77:042113, Apr 2008. DOI: 10.1103/PhysRevA.77.042113. 48, 128

[25] A. Cernoch, J. Soubusta, L. Bartusková, M. Dusek, and J. Fiurásek. Experimental realization of linear-optical partial swap gates. *Phys. Rev. Lett.*, 100:180501, May 2008. DOI: 10.1103/PhysRevLett.100.180501. 119, 121

[26] M.-D. Choi. Positive linear maps on c*-algebras. *Canadian Journal of Mathematics*, 24(3):520–529, 1972. DOI: 10.4153/CJM-1972-044-5. 45

[27] M.-D. Choi. Completely positive linear maps on complex matrices. *Linear Algebra and its Applications*, 10(3):285–290, 1975. DOI: 10.1016/0024-3795(75)90075-0. xiii, 45, 63, 64, 152

[28] C. Cohen-Tannoudji, B. Diu, and F. Laloe. *Quantum Mechanics*. Wiley, 1992. 9, 105

[29] C. Cohen-Tannoudji, J. Dupont-Roc, and G. Grynberg. *Photons and atoms: introduction to quantum electrodynamics*. Wiley professional paperback series. Wiley, 1997. 153, 154, 157

[30] J. M. Dominy, A. Shabani, and D. A. Lidar. A general framework for complete positivity. *ArXiv e-prints*, Dec. 2013. 58

[31] Y. Eidelman, V. Milman, and A. Tsolomitis. *Functional Analysis: An Introduction*. Graduate studies in mathematics, v. 66. American Math. Soc., 2004. 5

[32] K. Engel and R. Nagel. *One-Parameter Semigroups for Linear Evolution Equations*. Graduate Texts in Mathematics. Springer, 1999. 111

[33] F. Faisal. *Theory of Multiphoton Processes*. Advances in the Study of Communication and Affect. Springer, 1987. DOI: 10.1007/978-1-4899-1977-9. 155, 158

[34] A. Ferraro, L. Aolita, D. Cavalcanti, F. M. Cucchietti, and A. Acin. Almost all quantum states have nonclassical correlations. *Phys. Rev. A*, 81:052318, May 2010. DOI: 10.1103/PhysRevA.81.052318. 56

[35] R. Feynman, R. Leighton, and M. Sands. *Mainly mechanics, radiation, and heat*. The Feynman Lectures on Physics. Addison-Wesley, 1963. 80

[36] C. Fleming and B. Hu. Non-markovian dynamics of open quantum systems: Stochastic equations and their perturbative solutions. *Annals of Physics*, 327(4):1238–1276, 2012. DOI: 10.1016/j.aop.2011.12.006. 140

[37] J. Fraleigh and V. Katz. *A first course in abstract algebra*. Addison-Wesley world student series. Addison-Wesley, 2003. 10

[38] J. Frank. *Electron Tomography: Methods for Three-Dimensional Visualization of Structures in the Cell*. Springer, 2006. 23

[39] V. Gorini, A. Frigerio, M. Verri, A. Kossakowski, and E. Sudarshan. Properties of quantum markovian master equations. *Reports on Mathematical Physics*, 13(2):149–173, 1978. DOI: 10.1016/0034-4877(78)90050-2. xiii, 126, 127, 139

[40] V. Gorini, A. Kossakowski, and E. C. G. Sudarshan. Completely positive dynamical semigroups of n-level systems. *Journal of Mathematical Physics*, 17(5):821–825, 1976. DOI: 10.1063/1.522979. xiii, 126, 139

[41] T. F. Havel. Robust procedures for converting among lindblad, kraus and matrix representations of quantum dynamical semigroups. *Journal of Mathematical Physics*, 44(2):534–557, 2003. DOI: 10.1063/1.1518555. 34, 35, 63, 64, 65, 126, 128

[42] K.-E. Hellwig and K. Kraus. Pure operations and measurements. *Communications in Mathematical Physics*, 11(3):214–220, 1969. DOI: 10.1007/BF01645807. xiii

[43] H. F. Hofmann and S. Takeuchi. Quantum phase gate for photonic qubits using only beam splitters and postselection. *Phys. Rev. A*, 66:024308, Aug 2002. DOI: 10.1103/PhysRevA.66.024308. 122

[44] R. Horn and C. Johnson. *Matrix Analysis*. Cambridge University Press, 1990. 97

[45] M. Howard, J. Twamley, C. Wittmann, T. Gaebel, F. Jelezko, and J. Wrachtrup. Quantum process tomography and linblad estimation of a solid-state qubit. *New Journal of Physics*, 8(3):33, 2006. DOI: 10.1088/1367-2630/8/3/033. 116

[46] Z. Hradil, J. Rehacek, J. Fiurasek, and M. Jezek. 3 maximum-likelihood methods in quantum mechanics. In M. Paris and J. Rehacek, editors, *Quantum state estimation*, Lecture Notes in Physics, pages 59–112. Springer, 2004. DOI: 10.1007/b98673. 25, 65, 73

[47] K. Jacobs. *An Introduction to Quantum Measurement Theory*. (in preparation) http://www.quantum.umb.edu/Jacobs/books.html, 2013. 9

[48] T. F. Jordan, A. Shaji, and E. C. G. Sudarshan. Dynamics of initially entangled open quantum systems. *Phys. Rev. A*, 70:052110, Nov 2004. DOI: 10.1103/PhysRevA.70.052110. 13, 140

[49] K. Kieling, J. L. O'Brien, and J. Eisert. On photonic controlled phase gates. *New Journal of Physics*, 12(1):013003, 2010. DOI: 10.1088/1367-2630/12/1/013003. 122

[50] N. Kiesel, C. Schmid, U. Weber, R. Ursin, and H. Weinfurter. Linear optics controlled-phase gate made simple. *Phys. Rev. Lett.*, 95:210505, Nov 2005. DOI: 10.1103/PhysRevLett.95.210505. 122

[51] P. Kok and B. Lovett. *Introduction to Optical Quantum Information Processing*. Cambridge University Press, 2010. 95, 120, 158

[52] K. Kraus. General state changes in quantum theory. *Annals of Physics*, 64(2):311–335, 1971. DOI: 10.1016/0003-4916(71)90108-4. xiii

[53] K. Kraus. *States, Effects, and Operations, Vol. 190 of Lecture Notes in Physics*. Springer-Verlag, Berlin, 1983. xiii, 46, 50

[54] A.-m. Kuah, K. Modi, C. A. Rodriguez-Rosario, and E. C. G. Sudarshan. How state preparation can affect a quantum experiment: Quantum process tomography for open systems. *Phys. Rev. A*, 76:042113, Oct 2007. DOI: 10.1103/PhysRevA.76.042113. 103, 104, 117

[55] B. B. Laird, J. Budimir, and J. L. Skinner. Quantum-mechanical derivation of the bloch equations: Beyond the weak-coupling limit. *The Journal of Chemical Physics*, 94(6):4391–4404, 1991. DOI: 10.1063/1.460626. 127

[56] L. Landau and E. Lifshits. *Quantum Mechanics: Non-Relativistic Theory*. Teoreticheskai a fizika (Izd. 3-e) (Landau, L. D, 1908-1968). Pergamon Press, 1977. 4, 16, 83

[57] D. W. Leung. Choi's proof as a recipe for quantum process tomography. *Journal of Mathematical Physics*, 44(2):528–533, 2003. DOI: 10.1063/1.1518554. 63

[58] G. Lindblad. On the generators of quantum dynamical semigroups. *Communications in Mathematical Physics*, 48(2):119–130, 1976. DOI: 10.1007/BF01608499. xiii

[59] T. Lo and P. Inderwiesen. *Fundamentals of Seismic Tomography*. Geophysical Monograph Series, No 6. American Geophysical Union, 1994. 23

[60] R. Loudon. *The Quantum Theory of Light*. OUP Oxford, 2000. 95, 157, 158

[61] N. Luhmann. *Social Systems*. Stanford University Press, 1995. 3

[62] L. Mandel and E. Wolf. *Optical Coherence and Quantum Optics*. Cambridge University Press, 1995. DOI: 10.1017/CBO9781139644105. 95, 155, 157

[63] F. Masillo, G. Scolarici, and L. Solombrino. Some remarks on assignment maps. *Journal of Mathematical Physics*, 52(1):012101, 2011. DOI: 10.1063/1.3525832. 13, 128

[64] L. Mazzola, C. A. Rodriguez-Rosario, K. Modi, and M. Paternostro. Dynamical role of system-environment correlations in non-markovian dynamics. *Phys. Rev. A*, 86:010102, Jul 2012. DOI: 10.1103/PhysRevA.86.010102. xiv

[65] E. Merzbacher. *Quantum Mechanics*. Wiley, 1998. 16

[66] A. Messiah. *Quantum Mechanics:*. Dover books on physics. Dover, 1999. 16

[67] T. Mikio Nakahara and T. Ōmi. *Quantum Computing: From Linear Algebra To Physical Realizations.* Taylor & Francis Group, 2008. DOI: 10.1201/9781420012293. 95, 159

[68] M. Nielsen and I. Chuang. *Quantum Computation and Quantum Information: 10th Anniversary Edition.* Cambridge University Press, 2010. DOI: 10.1017/CBO9780511976667. 6, 9, 33, 39, 46, 73, 127

[69] H. Ollivier and W. H. Zurek. Quantum discord: a measure of the quantumness of correlations. *Physical review letters*, 88(1):17901, 2001. DOI: 10.1103/PhysRevLett.88.017901. 55

[70] M. Orszag. *Quantum Optics: Including Noise Reduction, Trapped Ions, Quantum Trajectories, and Decoherence.* Springer-Verlag Berlin Heidelberg 1997, 2008. 95

[71] M. G. Paris. The modern tools of quantum mechanics. *The European Physical Journal-Special Topics*, 203(1):61–86, 2012. DOI: 10.1140/epjst/e2012-01535-1. 9

[72] P. Pechukas. Reduced dynamics need not be completely positive. *Phys. Rev. Lett.*, 73:1060–1062, Aug 1994. DOI: 10.1103/PhysRevLett.73.1060. xiii, 10, 13

[73] P. Pechukas. Pechukas replies:. *Phys. Rev. Lett.*, 75:3021–3021, Oct 1995. DOI: 10.1103/PhysRevLett.75.3021. xiii, 10

[74] J. Preskill. *Quantum Computation Lecture Notes.* online notes: http://www.theory.cal tech.edu/~preskill/ph219/index.html#lecture, 2004. 46, 142

[75] T. C. Ralph and G. J. Pryde. Optical quantum computation. In E. Wolf, editor, *Progress in Optics*, volume 54 of *Progress in Optics*, pages 209–269. Elsevier, 2010. 120

[76] C. A. Rodriguez-Rosario. *The Theory of Non-Markovian Open Quantum Systems.* PhD thesis, University of Texas, 2008. 55, 56, 132, 140

[77] C. A. Rodriguez-Rosario, K. Modi, and A. Aspuru-Guzik. Linear assignment maps for correlated system-environment states. *Phys. Rev. A*, 81:012313, Jan 2010. DOI: 10.1103/PhysRevA.81.012313. 12, 13, 55, 103

[78] C. A. Rodriguez-Rosario, K. Modi, A. Kuah, A. Shaji, and E. C. G. Sudarshan. Completely positive maps and classical correlations. *Journal of Physics A: Mathematical and Theoretical*, 41(20):205301, 2008. DOI: 10.1088/1751-8113/41/20/205301. 55, 57, 139

[79] Rodriguez-Rosario, C. A., Modi, K., Mazzola, L., and Aspuru-Guzik, A. Unification of witnessing initial system-environment correlations and witnessing non-markovianity. *EPL*, 99(2):20010, 2012. DOI: 10.1209/0295-5075/99/20010. xiv

[80] A. Royer. Reduced dynamics with initial correlations, and time-dependent environment and hamiltonians. *Phys. Rev. Lett.*, 77:3272–3275, Oct 1996. DOI: 10.1103/PhysRevLett.77.3272. 142

[81] K. K. Sabapathy, J. S. Ivan, and S. G. nad R. Simon. Quantum discord plays no distinguished role in characterization of complete positivity: Robustness of the traditional scheme. *arXiv preprint arXiv:1304.4857*, 2013. 58

[82] J. Sakurai and J. Napolitano. *Modern Quantum Mechanics*. Addison-Wesley, 2010. 4, 17

[83] S. G. Schirmer and A. I. Solomon. Constraints on relaxation rates for n-level quantum systems. *Phys. Rev. A*, 70:022107, Aug 2004. DOI: 10.1103/PhysRevA.70.022107. 127

[84] F. Schwabl. *Quantum Mechanics*. Springer, 2007. 16, 155

[85] G. Sewell. *Quantum Mechanics and Its Emergent Macrophysics*. Princeton University Press, 2002. 139

[86] A. Shabani and D. A. Lidar. Maps for general open quantum systems and a theory of linear quantum error correction. *Phys. Rev. A*, 80:012309, Jul 2009. DOI: 10.1103/PhysRevA.80.012309. 127

[87] A. Shabani and D. A. Lidar. Vanishing quantum discord is necessary and sufficient for completely positive maps. *Physical review letters*, 102(10):100402, 2009. DOI: 10.1103/PhysRevLett.102.100402. 58

[88] A. Shaji. *Dynamics of initially entangled open quantum systems*. PhD thesis, University of Texas, 2005. 63, 113

[89] A. Shaji and E. Sudarshan. Who's afraid of not completely positive maps? *Physics Letters A*, 341:48–54, 2005. DOI: 10.1016/j.physleta.2005.04.029. xiii, 139

[90] C. Simon, V. Buzek, and N. Gisin. No-signaling condition and quantum dynamics. *Phys. Rev. Lett.*, 87:170405, Oct 2001. DOI: 10.1103/PhysRevLett.87.170405. 7

[91] C. Slichter. *Principles of Magnetic Resonance*. Springer Series in Solid-State Sciences. Springer, 1996. 126, 127

[92] P. Stelmachovic and V. Buzek. Dynamics of open quantum systems initially entangled with environment: Beyond the kraus representation. *Phys. Rev. A*, 64:062106, Nov 2001. DOI: 10.1103/PhysRevA.64.062106. 139

[93] W. F. Stinespring. Positive functions on c*-algebras. *Proceedings of the American Mathematical Society*, 6(2):pp. 211–216, 1955. DOI: 10.1090/S0002-9939-1955-0069403-4. xiii, 45

[94] E. C. G. Sudarshan, P. M. Mathews, and J. Rau. Stochastic dynamics of quantum-mechanical systems. *Phys. Rev.*, 121:920–924, Feb 1961. DOI: 10.1103/PhysRev.121.920. 1, 64

[95] D. Suter. *The Physics of Laser-Atom Interactions*. Cambridge Studies in Modern Optics. Cambridge University Press, 1997. DOI: 10.1017/CBO9780511524172. 95, 96

[96] J. Taylor. *An Introduction to Error Analysis: The Study of Uncertainties in Physical Measurements*. Physics - chemistry - engineering. University Science Books, 1997. 121

[97] D. Terno. Non completely positive maps in physics. *Conference on Quantum Information and Quantum Control II*, 2006. 139

[98] A. R. Usha Devi, A. K. Rajagopal, and Sudha. Open-system quantum dynamics with correlated initial states, not completely positive maps, and non-markovianity. *Phys. Rev. A*, 83:022109, Feb 2011. DOI: 10.1103/PhysRevA.83.022109. 104

[99] A. J. van Wonderen and K. Lendi. Reduced dynamics for entangled initial state of the full density operator. *Journal of Physics A: Mathematical and General*, 33(32):5757, 2000. DOI: 10.1088/0305-4470/33/32/311. 48

[100] J. von Neumann. *The Mathematical Foundations of Quantum Mechanics*. Felix Alcan, Paris, France, 1947. 2, 4

[101] Y. S. Weinstein, T. F. Havel, J. Emerson, N. Boulant, M. Saraceno, S. Lloyd, and D. G. Cory. Quantum process tomography of the quantum fourier transform. *The Journal of Chemical Physics*, 121(13):6117–6133, 2004. DOI: 10.1063/1.1785151. 125, 126, 133

[102] M. Wilde. *Quantum Information Theory*. Cambridge University Press, 2013. DOI: 10.1017/CBO9781139525343. 68

[103] C. Wood. Non-completely positive maps: properties and applications. *arXiv pre-print (quant-ph)*, arXiv:0911.3199, 2009. 115, 140

[104] H. P. Yuen. Uncertainty principle and the standard quantum limits. *arXiv pre-print (quant-ph)*, arXiv:quant-ph/0510069, 2005. 141

[105] N. Zettili. *Quantum Mechanics*. Wiley, 2009. 86

[106] W. H. Zurek. Einselection and decoherence from an information theory perspective. *Quantum Communication, Computing, and Measurement 3*, pages 115–125, 2002. DOI: 10.1007/0-306-47114-0_17. 55

[107] W. H. Zurek. Decoherence, einselection, and the quantum origins of the classical. *Rev. Mod. Phys.*, 75:715–775, May 2003. DOI: 10.1103/RevModPhys.75.715. 3, 55

[108] K. Zyczkowski and I. Bengtsson.   On duality between quantum maps and quantum states.   *Open systems & information dynamics*, 11(01):3–42, 2004. DOI: 10.1023/B:OPSY.0000024753.05661.c2. 139

Author's Biography

JAMES M. MCCRACKEN

James M. McCracken received his Physics M.S. studying decoherence in quantum dot qubit structures at the University of Central Florida, and he received two B.S. degrees, one in Physics and one in Astrophysics, from the Florida Institute of Technology. He was a member of both the theoretical quantum computing team at Science Applications International Corporation (SAIC) and the informatics group at the U. S. Naval Research Laboratory in Washington, DC. James is currently a Ph.D. candidate studying time series causality and geomagnetic storm prediction at George Mason University in Fairfax, Virginia.